中国云南省高黎贡山南段
China: Yunnan,
Southern Gaoligongshan

Douglas F. Stotz, Emily J. Harris, Debra K. Moskovits,
Ken Hao (郝光明), Shaoliang Yi (易绍良), and
Gerald W. Adelmann, 主编 editors

2003年2月 FEBRUARY 2003

参加单位 Participating Institutions:

美国菲尔德博物馆
美国美中艺术交流中心
中国西南林学院
中国高黎贡山国家级自然保护区保山管理局
美国开阔地计划
中国云南省对外文化交流协会美国
建筑设计公司

The Field Museum
Center for United States-China Arts Exchange
Southwest Forestry College (Kunming)
Gaoligongshan Nature Reserve Management Bureau
Openlands Project
Yunnan Provincial Association for Cultural Exchanges
    with Foreign Countries
Skidmore, Owings & Merrill, LLP

协助考察 Collaborators:

中国云南省保山市政府
中国云南省保山市百花岭村民委员会

Baihualing Village Association – China
Municipal Government of Baoshan City – China

出版 RAPID BIOLOGICAL INVENTORIES REPORTS ARE PUBLISHED BY

菲尔德博物馆
环境与保护项目
美国芝加哥市南湖滨大道1400号
邮编: 60605-2496
电话: 312.665.7430
传真: 312.665.7433
网址: *www.fieldmuseum.org*

THE FIELD MUSEUM
Environmental and Conservation Programs
1400 South Lake Shore Drive
Chicago, Illinois 60605-2496 USA
T 312.665.7430, F 312.665.7433
www.fieldmuseum.org

主编 **Editors:** Douglas F. Stotz, Emily J. Harris,
Debra K. Moskovits, 郝光明 Ken Hao, 易绍良 Shaoliang Yi,
and Gerald W. Adelmann

装帧设计 **Design:** 公司 (美国芝加哥市)
Costello Communications, Chicago

中文翻译 **Translations:** 易绍良 Shaoliang Yi, 何新元 He Xinyuan,
皮 英 Pi Ying, 文 军 Jun Wen, and 郝光明 Ken Hao

建议引用: Stotz, D.F., E.J. Harris, D.K. Moskovits, 郝光明, 易绍良,
and G. W. Adelmann (主编). 2003. 快速生物资源调查报告 (4)
(中国云南省高黎贡山南段). 美国伊利诺依州芝加哥市:菲尔德博
物馆 (The Field Museum) 出版社.

Suggested Citation: Stotz, D.F., E.J. Harris, D.K. Moskovits,
K. Hao, S. Yi, and G. W. Adelmann (eds.). 2003. China: Yunnan,
Southern Gaoligongshan. Rapid Biological Inventories Report
No. 4. Chicago, Illinois: The Field Museum.

封面照片 **Cover Photograph:** 滑鼠蛇 (*Ptyas mucosus*) — 一种
大型、无毒蛇, 在南亚广泛分布, 照片提供 H. Bradley Shaffer.
封里照片 高黎贡山西坡大塘附近稻田, 照片提供 文军 J. Wen.
Ptyas mucosus—a large, harmless, and widespread snake in
southern Asia, by H. Bradley Shaffer. Inner-cover photograph:
An agricultural valley near Datang at the western base of the
Gaoligongshan Mountains, by Jun Wen.

彩色插图照片 **Color Plate Photography:** 图 Figures 2B, 2D,
2E, 5A-F, 6A-E, 8B, 8D, H. B. Shaffer; 图 Figures 1, 2A, 2C,
2F, 3A-D, 4A-E, 4G, R. Foster; 图 Figure 3E, D. Moskovits;
图 Figure 4F, 文军 J. Wen; 图 Figures 7A-E, G. Mueller; 图 Figures
8A, 8C, 8F, 8G, 9A-F, A. Thies; 图 Figure 8E, A. Underhill;
图 Figure 2, NASA; 图 Figure 3, 高黎贡山国家级自然保护区保山
管理局 Gaoligongshan Nature Reserve Management Bureau
(Baoshan); 地图114页 map on page 114, J. Seagard.

 本报告系利用回收纸印刷 Printed on recycled paper.

目录

野外考察小组成员

Gerald W. Adelmann（组织者）
开阔地计划（Openlands Project）
美国伊利诺依州芝加哥市
jadelmann@openlands.org

艾怀森（兽类调查小组）
中国高黎贡山国家级自然保护区保山管理局
GLGS@BS.YN.cninfo.net

杨斌（真菌调查小组）
中国西南林学院
yangbinyb@hotmail.com

邓莉兰（植物调查小组）
中国西南林学院
gsfan@public.km.yn.cn

Victoria Drake（社会文化调查小组）
开阔地计划（Openlands Project）
美国伊利诺依州芝加哥市，及美国菲尔德博物馆
jecvcd@earthlink.net

Robin B. Foster（生态调查小组）
美国菲尔德博物馆环境与保护项目
rfoster@fieldmuseum.org

葛尚义（真菌调查小组）
中国高黎贡山国家级自然保护区保山管理局
GLGS@BS.YN.cninfo.net

郝光明（组织者）
美国哥伦比亚大学美中艺术交流中心
kmh101@columbia.edu

Peter J. Kindel（社会文化调查小组）
美国Skidmore, Owings & Merrill建筑设计公司
peter.j.kindel@som.com

李正波（组织者，鸟类调查小组）
中国高黎贡山国家级自然保护区保山管理局
GLGS@BS.YN.cninfo.net

蔺如涛（兽类调查小组）
中国高黎贡山国家级自然保护区保山管理局
GLGS@BS.YN.cninfo.net

孟世良（生态调查小组）
中国高黎贡山国家级自然保护区保山管理局
GLGS@BS.YN.cninfo.net

Debra K. Moskovits（组织者，鸟类调查小组）
美国菲尔德博物馆环境与保护项目
dmoskovits@fieldmuseum.org

Gregory M. Mueller（真菌调查小组）
美国菲尔德博物馆植物学部
gmueller@fieldmuseum.org

覃家理（生态调查小组）
中国西南林学院
qinjiali@lol365.com

权锐昌（鸟类调查小组）
中国科学院昆明动物研究所
quanre@163.com

H. Bradley Shaffer（两栖与爬行动物调查小组）
美国加里弗尼亚大学（Davis分校）
hbshaffer@ucdavis.edu

施晓春（植物调查小组）
中国高黎贡山国家级自然保护区保山管理局
GLGS@BS.YN.cninfo.net

Douglas Stotz（鸟类调查小组）
美国菲尔德博物馆环境与保护项目
dstotz@fieldmuseum.org

Adam Thies（生态旅舍和游客接待中心设计）
美国Skidmore, Owings & Merrill建筑设计公司
adam.thies@som.com

Anne Underhill（社会文化调查小组）
美国菲尔德博物馆人类学部
auhill@fieldmuseum.org

王天灿（两栖与爬行动物调查小组）
中国高黎贡山国家级自然保护区保山管理局
GLGS@BS.YN.cninfo.net

文军（植物调查小组）
美国菲尔德博物馆植物学部
jwen@fieldmuseum.org

杨建美
中国昆明云南师范大学旅游规划研究中心

易绍良（组织者）
中国西南林学院
a1234567@public.km.yn.cn

张　宇（两栖与爬行动物调查小组）
中国西南林学院
swfczhangy@163.com

朱明育（社会文化调查小组）
中国高黎贡山国家级自然保护区保山管理局
GLGS@BS.YN.cninfo.net

赵晓东（组织者）
中国高黎贡山国家级自然保护区保山管理局
GLGS@BS.YN.cninfo.net

协助单位

中国云南省保山市百花岭村民委员会
中国云南省保山市政府

合作机构介绍

美国菲尔德博物馆

美国菲尔德博物馆 (The Field Museum) 是一个以收藏为
基础的、致力于自然和文化多样性研究和教育的机构。
博物馆拥有一大批人类学、植物学、地质学、动物学和
保护生物学等学科的专家，开展诸如进化、环境生物学
和文化人类学等方面的研究工作。环境与保护项目
(Environmental and Conservation Programs 'ECP')
是博物馆的一个分支机构，主要致力于将科学知识运用
于持久的保护行动。随着世界范围内的自然生物多样性
的加速丧失，ECP的使命是将博物馆的资源 (科学知识、
收藏、新颖的教育活动等) 尽量应用到地区、国家和国
际范围的保护活动。

联系：
The Field Museum
1400 S. Lake Shore Drive
Chicago, Illinois 60605-2496 U.S.A.
312.922.9410 tel
www.fieldmuseum.org

美中艺术交流中心

哥伦比亚大学美中艺术交流中心 (Center for
United States-China Arts Exchange) 为美中关系正常化
以前由周文中教授，哥伦比亚大学弗里茨 雷纳
(Fritz Reiner)荣誉教授创建的。1979年以来，交流中心
在艺术教育和自然保护等领域开展了许多重大的项目，
这些项目为整个亚太地区以及欧洲和美国的机构提供了
合作的机会。1990年以来，中心一直将工作重点放在
中国云南省的一个项目上，该项目涉及环境及文化等领
域，以当地少数民族为目标群体，目的是帮助这些少数
民族制定出一个综合性的、保存传统文化和保护生态环
境的策略。目前已经有几百名来自中国、美国、欧洲和
亚洲的专家学者参与了此项目，同时项目还调动了数以
千计的当地文化和环境保护人员的参与。

联系：
Center for United States-China Arts Exchange
423 West 118th Street, #1E
New York, New York, 10027 U.S.A
us_china_arts@yahoo.com

西南林学院

西南林学院位于中国云南省昆明市，属于中国国家林业局和云南省政府共建学校。西南林学院为中国西部地区唯一一所独立办学的高等林业院校。学校的主要任务是培养林业高级人才，开展林业研究和为中国特别是西南地区的林业生产提供技术服务。学校占地100公顷左右，2002年在校学生约6，000人（少数民族学生占学生总数的20%左右）。学校目前有13个教学分院和系部，10余个主要研究机构。学校的主要研究领域包括林学、资源管理、乡村规划、野生动物与自然保护区管理、水土保持、生态旅游和木材加工等。多年来，西南林学院承担和参与了云南省许多自然保护区的建立和升级的综合考察工作。

联系：
中国云南省昆明市白龙寺
邮编： 650224
电话：+86-871-386-3211
传真：+86-871-386-3217
oicswfc@public.km.yn.cn
http://www.swfc.edu.cn

高黎贡山国家级自然保护区保山管理局

成立于1993年，管理着整个保护区约25%（99,675公顷）的面积，其主要职责为：（1）保护区管理，包括执法等；（2）与其它机构一起开展教育与研究活动；（3）开展可持续的生态旅游活动。保护区下属两个管理所（隆阳管理所和腾冲管理所）、11个管理站和两个林业派出所。管理局人员包括管理人员、技术人员和护林员等。

联系：
中国云南省保山市公园路2号
邮编： 678000
电话：+86-875-2121858
传真：+86-871-2120288
GLGS@BS.YN.cninfo.net

美国开阔地计划

开阔地计划(Openlands Project)是一个私营的、非赢利性机构，成立于1963年。机构的宗旨是保护和扩大现有开阔空间（土地和水体）并加强其功能，以便为芝加哥市区的所有居民提供卫生、自然的环境和更加适合居住的生活空间。自成立以来，该机构已经参与保护了50,000多英亩的公园、自然生境、保护区、自行车旅游便道、湿地、城市花园和绿地等。1982年，开阔地计划成立了运河走廊协会，该协会领导建立了美国第一个国家遗产走廊带。开阔地计划总裁杰瑞德.艾德曼先生也是哥伦比亚大学美中艺术交流中心董事会成员，是他将菲尔德博物馆的专家和Skidmore Owings & Merrill建筑设计公司的专家们介绍到高黎贡山地区的。

联系:
Openlands Project
25 East Washington, Suite 1650
Chicago, Illinois 60602 U.S.A
312.427.4256 tel
info@openlands.org

云南省对外文化交流协会

云南省对外文化交流协会是云南省最大的、从事对外交流的专业组织。协会的目标是向国外介绍云南、加强云南与各国人民的友谊和通过开展合作项目和互访等方式与各国人民交流知识和经验。

联系
中国云南省昆明市弥勒寺
邮编： 650021

Skidmore, Owings & Merrill LLP (SOM)
建筑设计公司

SOM建筑设计公司成立于1936，是世界上从事建筑、
城市设计、工程和室内装饰方面顶尖设计公司之一。
自成立以来，公司已经在全世界的50多个国家完成了
10,000多个设计项目。公司还获得过800多项奖项，
1961年，公司荣获美国建筑师院颁发的一等奖。
作为环境设计圈内的领头羊，SOM致力于为商业和非
商业性客户设计标志性建筑。 通过与政府机构和非赢
利性机构合作，SOM已经为包括云南项目在内的许多
国内和国际机构提供了义务性的设计和技术服务

联系:
Skidmore, Owings & Merrill, LLP
224 South Michigan Avenue, #1000
Chicago, Illinois 60604 U.S.A.
www.som.com

本报告是哥伦比亚大学美中艺术交流中心"云南倡议"的一个示范项目的总结。
"云南倡议"的中国专家委员会将众多国外专家带到云南与当地同行、领导和群众合作为保护云南省的生态环境和文化资源出谋划策。

"云南倡议"的指导原则

保护　　　　应该在保护文化、生态和社会的前提下谋求发展。

包容　　　　发展和保护策略一定要考虑所有的民族、建立在当地的文化传
　　　　　　统上，要与云南省建设民族文化大省的发展战略一致。

公众教育　　要保证致力于可持续发展的项目的长期成功，必须要让公众充
　　　　　　分认识到文化和环境的价值。

旅游　　　　作为经济发展的潜在推动力，旅游发展必须以弘扬云南省的民
　　　　　　族文化和加强云南省的生态基础为前提，同时还必须要能够给
　　　　　　当地人带来直接的社会经济效益。

合作　　　　保护与发展协调发展的策略必须建立在地方、国内和国际合作
　　　　　　的基础上。

2002年6-7月间，由自然科学家、建筑师、规划师和文化旅游专家等组成的专家小组开展了野外考察。本报告即是对本次跨学科考察的总结。报告中的建议部分就是本着上述原则提出的。此外，本报告还融入了美中艺术交流中心2001年在魏山县考察的报告内容。

致　　谢

自1990年以来，美中艺术交流中心的创始人兼主任周文中教授，哥伦比亚大学的弗里茨.雷纳 (Fritz Reiner) 荣誉作曲教授，一直不遗余力地关心中国云南省的民族文化以及环境保护事业。周文中教授卓越的领导才能为此次生物资源和社会资源的快速调查的顺利开展奠定了坚实的基础，同时也为今后在云南省进行环境保护和振兴当地资源的工作铺平了道路。

与周文中教授一道致力于云南环境保护行动的还有云南社会科学联盟的主任顾伯平先生。顾先生不懈的努力敲开了云南参与国际合作的大门。在此，我们还要感谢云南省委办公厅的范建华教授。范教授精力充沛，对云南的风土人情、历史和文化有独到的见解。这些因素使得范教授成为云南环境保护行动成功实施的关键之一。

实施高黎贡山国家级自然保护区生物资源和社会资源快速调查的各小组还得到了云南省保山市委宣传部部长杨文虎先生的大力支持。杨文虎先生组织并参加了"高黎贡山生物资源和社会资源快速调查成果新闻发布会"。

对此次调查作出贡献的还有西南林学院副院长杨宇明教授。杨宇明教授对此次调查提供了植物学知识方面的宝贵支持。他对国际文化和科技合作项目的见解使之成为领导中方人员的最佳人选。参加此次调查的中方人员中，有不少人是杨宇明教授的同事和研究生。

西南林学院的吴德友教授承担了西南林学院快速调查小组的组织和准备工作，为调查提供了必要的保障。吴教授细心周到的安排和准备工作使得艰苦的野外协作调查工作得以按计划顺利开展。

中科院西双版纳热带植物园的副主任郭辉军教授是云南省国际知名的保护和植物学专家。他给参加此次调查的美方小组进行了云南省和高黎贡山相关知识的介绍，为此次调查提供了宝贵的学术支持。

在此，我们还要特别感谢高黎贡山国家级自然保护区管理局副局长李正波先生。李正波先生为此次调查的顺利进行提供了坚实的后勤保障。同时我们还感谢李正波先生以及保护区的其他工作人员为我们提供了当地大量的相关信息。

所有的中方科学家以及调查人员对不懂中文的美方人员表示了充分的理解和支持。特别需要感谢的是文军博士，她的翻译为美方人员解决了不少因语言障碍而导致的问题。

菲尔德博物馆得以参加到云南环境保护行动中来，完全要归功于John W. McCarter Jr. 先生的努力和领导。为促成此次调查的成功实施，他对云南进行了一次先期的访问。今后，John W. McCarter Jr. 先生将会不遗余力地支持我们的工作。

菲尔德博物馆文化中心主任Alaka Wali女士为社会资源的调查工作提供了资金和技术保障。她发明了一套当地社会资源调查和保护的方法。同时她还为白花岭村调查小组提供了她在南美洲以及美国伊利诺斯州开展类似项目的成功经验。

菲尔德博物馆的Robin Groesbeck先生为举办相关的展览提出了宝贵的建议。博物馆研究员Ben Bronson 先生也为举办展览及开展相关的研究提出了宝贵的建议此外，遗产、旅游和交流集团(HTC) 的国际文化旅游专家Cheryl Hargrove也为调查小组提供了指导和建议，使此次调查活动得以顺利实施。

在Philip Enquist先生强有力的支持和领导下，AIA, Skidmore Owings & Merrill建筑设计公司提供了建设生态旅游旅舍及游客中心的设计方案。

以下机构为此次调查活动提供了必要的资金支持，它们是: John D. and Catherine T. MacArthur Foundation 麦克阿瑟基金会、美中艺术交流中心和菲尔德博物馆。在此，我们再次表示感谢。

报 告 概 要

野外考察日期	生物调查：2002年6月17日-26日；社会调查：2002年7月1日-14日
考察地点	考察地点为位于中缅边境地区的高黎贡山国家级自然保护区南段（云南省保山地区）。具体考察地点为：(1)高黎贡山东坡从百花岭管理站（海拔约1,500m左右）沿古丝绸之道一直到南斋公房垭口（海拔3,100m左右）；(2)高黎贡山西坡的大塘，考察范围从1,850m到2,700m；(3)赧亢，位于保护区南端垭口处，海拔2,000m至2,200m之间（图 3）。
考察社区	社会、文化调查小组一共对百花岭行政村的8个自然村（汉龙、大鱼塘（上社和下社）、帮迈-古兴寨、桃园、老蒙寨-百花岭-麻栎山、芒岗和芒晃）（图3）。这些村子都与保护区的边界相邻，从东面进入保护区都必须从这些村子过。
考察的生物类群	高等植物、大型真菌、两栖与爬行动物、鸟类、大型哺乳动物
调查结果要点	高黎贡山地区是一个生态特征独特、南北、东西及温带和亚热带动植物区系大交汇的地方。该地区从东到西、从低海拔河谷到高海拔的山峰都分布着连绵不断的森林。这些森林植被孕育着纷繁复杂的生态群落。分布在高黎贡山山脉的众多高山峡谷、峻岭溪流为新物种的形成创造了得天独厚的条件，同时，当地常年湿润的气候等又为防止古老物种的消失创造了难得的环境。在不到10天的野外考察中，生物考察组的科学家门充分领略了保护区内极其丰富的动植物区系，发现了一些新种和新记录。但是，在保护区边界以外，森林已经遭到了严重的破坏。 居住在高黎山脚下的众多村子反映了云南省丰富的文化多样性。在社会文化小组为期14天的野外考察中，专家们对保护区东坡周边地区的村寨的资源和能力进行了考察和评估。这些资源将是在当地社区开展文化和生态兼容的经济活动如生态旅游活动的切入点。 兹将此次考察的结果要点总结如下： **社会资源和能力** 社会文化调查主要在百花岭行政村进行，该村为保护区周边地区109个行政村之一（保山管理局，2002）。此前，在该地区尚未开展过这类调查。百花岭行政村常驻人口约2,100人，450户，主要有汉、傈僳、白、傣、彝和回等民族。在中国的55个民族中，25个分布在云南省（云南省人口普查办公室，1992）。与云南其它地方的民族相比，这里的民族通过服饰、手工艺品、语言、节日和生活习惯来表达其民族身份的现象没有那么明显。这些村子的经济还完全是农业型经济，村民的主要（85%以上）收入来源于种植甘蔗、水稻和咖啡。生产生活用燃料是村民最关心的问题之一。全村90%以上的农户使用了节柴灶，使能源消耗减少了60%。这些村子无不具有鲜明的特色。文化资源有从古代丝绸之路一直到第二次世界大战等不同时期的历史遗迹。村民们大多能够生产带有民族特色的传统手工艺品如绣花鞋（图8E）和藤条用具等和制作传统食品。村子里已经有一些自发性的群众组织如高黎贡山农民生物多样性保护协会等。这些组织通过示范户在村子里实施了一些示范项目。在大鱼塘有一个规模较小的旅游协会；芒晃村有一个妇女协会。村民们普遍十分重视教育。

植被和植物区系

在为期9天的短暂考察中，我们一共观察到了1,000多种高等植物（包括6个新记录），采集到300种植物标本，并对250多种植物进行了拍照（图4）。植物区系中有10%左右属于高黎贡山地区特有种。在3个考察地点之间，植物区系存在着很大差异。由于地质条件的差异，东（百花岭一线）西坡（大塘附近）植物区系也明显不同。我们至少发现了3个植物新种（葡萄科1种，五加科2种）。保护区的蕨类植物非常丰富（附录2），考察中发现了许多属和种的新记录。

大型真菌

真菌考察小组在考察过程中共发现了200多种大型真菌（考察范围在海拔1,500m到2,400m之间），并采集了150种标本（附录1）。在记录到的种类中，只有22中被列入了先前的高黎贡山真菌名录。考察中发现高黎贡山汇集了许多北温带类型、热带亚洲种类和中国特有种，还发现了许多北美东部和东亚间断分布种类。大型真菌对于维持高质量的自然群落十分重要，同时也是当地群众重要的食物和经济来源。

两栖、爬行动物

两栖爬行动物考察小组一共发现了1种蛇类、4种蜥蜴、1种蝾螈和15-21种(视进一步鉴定结果，见图5及附录3)蛙类。在这些发现中，有三种蛇类和两种蛙类属于该地区的新记录，还有一种高山蛙类属高黎贡山特有种。保护区内的普通两栖类的多度是大塘附近农田中的多度的2-10倍(图3D)。保护区周围地区农业生产中大量使用农药等化学物质对蛙类具有非常有害的影响。

鸟类

在9天的考察中，鸟类考察小组在保护区内一共记录到179种鸟类（附录4），包括23种新记录。高黎贡山，特别是海拔2,400m以下的地区，森林鸟类非常丰富。鸟类在海拔高度上的物种替代率比较高，但是，在考察中没有发现明显的鸟类群落随海拔高度变化的情况。在我们观察到的鸟类中有43种（25%）属于分布范围狭窄种类，包括云南省山地特有种和东喜玛拉雅种类。保护区目前的鸟类名录上至少有19种处于"受威胁"或"接近受威胁"状况。

大型兽类

兽类调查小组在考察中共发现了42种大型动物(附录5)的直接或间接证据，其中有13种为国家级保护种类（包括4种灵长类动物和小熊猫）。本次调查中比较重要的发现就是在海拔2,000m附近地区有小熊猫（*Ailurus fulgens*）的广泛分布，而过去认为该物种一般都分在海拔3,000m左右的高山地带。

主要威胁因子

高黎贡山国家级自然保护区面临的直接的威胁因子主要有：(1) 农业生产（尤其是在保护区下部边沿地带，农药及农用化学物质的使用 、农业活动对溪流、水塘等的破坏和污染物的漂移等）；(2) 在保护区内开荒、建牧场和林下放牧等活动；(3) 由于缺乏村民支付得起的替代能源，当地社区需要 薪材作为燃料。由于信息闭塞，当地群众缺乏环境友好的措施来取代目前的农业活动，所以，这些对环境具有破坏作用的农业生产方式还有可能继续下去。低海拔地区森林被毁，使当地极其丰富的动植物资源(有些还是当地仅有的)濒临灭绝。同时,低海拔地区(主要在保护区外)动植物资源的消失必然会影响到高海拔地区(保护区内)的生物群落的动态平衡。最后,虽然生态旅游活动等能够对当地带来潜在的经济利益和发展机遇,但是如果开发和管理不当,不对当地脆弱的生物群落和人类文化进行严加保护的话,旅游活动就会威胁到保护区的完整性。

现 状	高黎贡山国家级自然保护区保护着高黎贡山海拔较高地段的405,549公顷地区。保护区下部边缘的海拔在1,500m到2,500m之间。海拔比较高的地区为保护区的核心区,禁止任何人为活动(图3)。不过,古丝绸之路一线未被划为核心区,游客可沿着古丝绸之路一直到达南斋公房垭口海拔3,100m左右的地方。保护区边界以下的地方主要是一些零散的庄稼地、牧场和受到严重干扰的林子(图2B, 2E, 3C, 3E),没有纳入正规的保护计划。1994年,中国林业部批准允许在保护区核心区以外的8,550公顷(占总面积的6.8%)范围内开展旅游活动。

关于保护与管理方面的主要建议

1) **延伸保护管理活动的范围,使从怒江到龙江边广大范围内都能得到有效地保护 (图3)**
   尽量将保护区的边界向下扩展到低海拔地区。低海拔山坡上的片片零星的森林植被正在被逐渐开垦为农地(图2B, 2C, 3D),使这些地方丰富多彩的动植物资源濒临灭绝。通过与周边社区合作在保护区以外的地区开展生态友好型经济活动能够将有效保护区域从怒江边一直延伸到龙江边,从而使高低海拔的动植物资源都能得到保护。

2) **核心区严禁开展任何活动,开放部分区域用于科学研究;尽可能将核心区延伸到低海拔地区**
   鉴于高黎贡山森林的极其重要的生物学价值和周边地区人口对保护区施加的巨大压力,我们建议严禁在现在的核心区范围(主要是高海拔地区,见图3)开展任何形式的人为活动,并尽可能将核心区向低海拔地区延伸。

3) **恢复和保护周边低海拔地区的森林植被;通过植被恢复等措施将目前分散的林块联结起来,并在大的保护区域之间建立走廊带**
   为了保护低海拔地区这些具有全球重要意义的生物资源,我们建议保护区管理部门与周边地区村民密切合作,共同恢复社区周围尚未被完全破坏的森林植被(图3C),并利用乡土树种在宜林荒山开展造林活动以扩大和联结动植物的有效生境。

4) **加强百花岭村参与规划和实施生态旅游活动的能力**
   保证当地居民能够真正参与旅游活动的规划并从旅游开发中获益的一种方式就是成立一个由所有自然村的代表参加的、积极开展活动的村级生态旅游协会(类似于农民生物多样性保护协会)。该协会可以与村委会一道讨论和实施旅游计划和规章制度等。目前大鱼塘的旅游管理委员会的规模和覆盖范围都还太小,不能代表所有自然村的利益,所以还无法在更大的范围内起作用。

5) **在低海拔地区的山坡和谷地试验和实施一些有利于生态保护的农业生产活动和生态恢复活动**
鼓励农民发展多种经营，减少对污染环境的（有时也是很贵的）化肥和农药的使用。

6) **开发适合当地社区支付能力的替代能源**
目前，周边社区主要使用薪材作为燃料。开展旅游活动必然会增加对燃料的需求，从而增加保护区资源的压力。目前有关部门正在该地区推广使用一些替代能源如沼气等，但是当地村民没钱修建沼气池及购买配套设施。

关于生态旅游方面的建议	1) **保证所有生态旅游活动能够给保护区和周边村民带来直接的利益** 2) **对保护区的游客承载量进行研究，并根据研究结果控制游客接待数量** （见附录8） 3) **旅游活动的开展和旅游基础设施的修建要尽量减少对敏感生物群落的影响，并要有利于加强周边村寨的能力；将基础设施修建在保护区外面** 充分利用现有基础设施，修建入口处旅舍和游客接待中心，作为旅游管理中心。建筑风格要符合当地传统（见附录8） 4) **限制和监测可能对生物群落造成损害的活动** 主要方法包括 (i) 垃圾废物合理处理; (ii) 不用木材作燃料做饭和取暖; (iii) 游客应在预先设计的林间小道范围内活动（小道应进行合理管理以减少土壤流失); (iv) 尽量减少用牲畜作运输工具; (v) 控制露宿量。 5) **将高黎贡山国家级自然保护区与滇西众多旅游景点统一开发** 这样既可以避免游客都集中在保护区、增加保护区的的压力，又能丰富游客的旅游体验，同时还有利于提高社区能力和保护本地文化和景观
长期的保护目标	1) 使高黎贡山国家级自然保护区成为一个具有全球重要性的保护区—从海拔1,500m左右的缓坡地带到 海拔4,000m以上的峻峭山峰—保护着来自不同动植物区系的生物群落。 2) 使目前受到威胁的低海拔地区的丰富的动植物资源的生境得到恢复，这些动植物资源中有很多是本地区特有的；珍贵的自然资源得到持续管理。 3) 使保护区成为政府与当地村民合作、开展有利于生态和文化保护、能给保护区和当地社区都带来利益的生态旅游的典范；国内外游客到该地区开展生态旅游活动，领略和欣赏当地丰富的生物和文化资源。 4) 使保护区和周边地区得到综合管理，社区开始采用有利于流域保护的农业措施，并减少对环境有破坏作用的农业化学物的使用。

为何选择高黎贡山开展此次考察？

高黎贡山位于中国西南边陲、中缅边境，是东与西、南与北，以及温带和亚热带动植物区系的交汇处，汇聚了独特的生物群落和生物地理成分。从东到西、从低到高的连绵森林为属于喜玛拉雅界、古北界和东南亚东洋界热带成分等的动植物区系成分提供了通道。高黎贡山具有世界上得天独厚的条件，在这里可以看到从热带到温带森林植被的各种类型。该地区生物多样性和特有种非常集中，也是保护地球上丰富的生物资源的优先区域。

高黎贡山同时也是各种文化交融、碰撞的地方和各种历史事件的发生地。自古以来，该地区的主要河流如怒江和龙川江流域就开始有人类居住。从公元前四世纪以来，位于高黎贡山山脉南段的古丝绸之路就一直是中国中原通往印度、阿富汗、巴基斯坦等国的重要商业、贸易和文化通道。此次我们文化调查的重点是与高黎贡山国家级自然保护区边界相连的百花岭行政村（图3）。该村有8个自然村，约450户人家。这些村子居住着汉、白、傈僳、彝、回和傣族等多个民族，充分体现了云南文化多样性的特点。

虽然该保护区早在1986年就被中国国务院批准为国家级自然保护区，在2000年又被联合国教科书组织接纳为人与生物圈保护区，但是，高黎贡山山脉壮观的环境还在继续遭受着巨大的压力。在低海拔地区保护区外的山坡上蕴藏着极其丰富的生物多样性，但是随着这些山坡上的森林的迅速消失，这些生物多样性受到的威胁也越来越大。中国政府批准在保护区的一部分开展生态旅游活动。高黎贡山的自然瑰宝的长期存在有赖于当地社区能否全面参与保护区的管理和保护。要保护高黎贡山这些世界上独一无二的自然宝藏，在开展经济活动时，必须考虑这些经济活动对于生态和文化保护是否有利，是否能够给当地群众的发展和当地的生物群落的保护带来直接的利益。

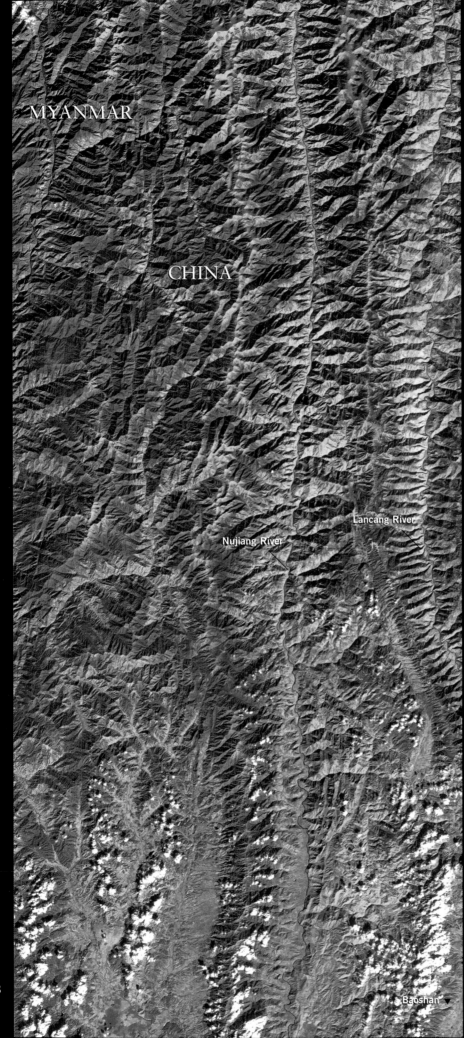

MYANMAR

CHINA

Lancang River

Nujiang River

Baoshan

CHINA: YUNNAN

图 FIG. 2 峻峭崎岖的喜玛拉雅山脉的高山峡谷中奔腾着亚洲许多大江大河。图中位于中缅边境地区云南一侧的河流就是怒江（萨尔温江）和澜沧江（湄公河），两条大江并行而下。高黎贡山山脉逶迤于怒江西岸。高黎贡山国家级自然保护区位于本卫星图像中的南端，保山以西。The craggy Himalayan range channels Asia's great rivers through narrow gorges. Seen here close to the Yunnan border with Myanmar are the Nujiang (Salween) and the Lancang (Mekong) Rivers, running side-by-side. The Gaoligong Mountains stretch along the western bank of the Nujiang River. The National Nature Reserve is at the southern edge of this NASA/LANDSAT image, west of Baoshan.

50 km   100 km

图 FIG. 2A 高黎贡山有很多深切的峡谷。图中瀑布位于百花岭管理站附近，是一处壮观的旅游景点。Deep canyons dissect the Gaoligong mountains. This waterfall, close to the Baihualing Station, is a magnificent tourist attraction.

图 FIG. 2B 在大塘附近河谷地带（海拔 1,850 m），森林已经破坏殆尽，只剩下些零星林块。Only small pockets of forest survive in the valley floor (1,850 m) around Datang.

图 FIG. 2C 旱田及其它农作物正在蚕食低海拔地区的森林。这些森林蕴藏着丰富的物种。Dryland rice fields and other crops threaten the lower-elevation forests, which are rich in species.

图 FIG. 2D 在高海拔山坡上，森林繁茂，保存完好。On the higher slopes, the protected forests thrive largely undisturbed.

图 FIG. 2E 高海拔地区快速流动的河流为许多特布的蛙类和鸟类提供了关键生境。Fast-flowing, high-elevation streams provide crucial habitat for specialized species of frogs and birds.

图 FIG. 2F 竹子构成了高海拔地区许多森林类型的主要林下植被。Bamboo dominates the understory in several high-elevation forests.

19

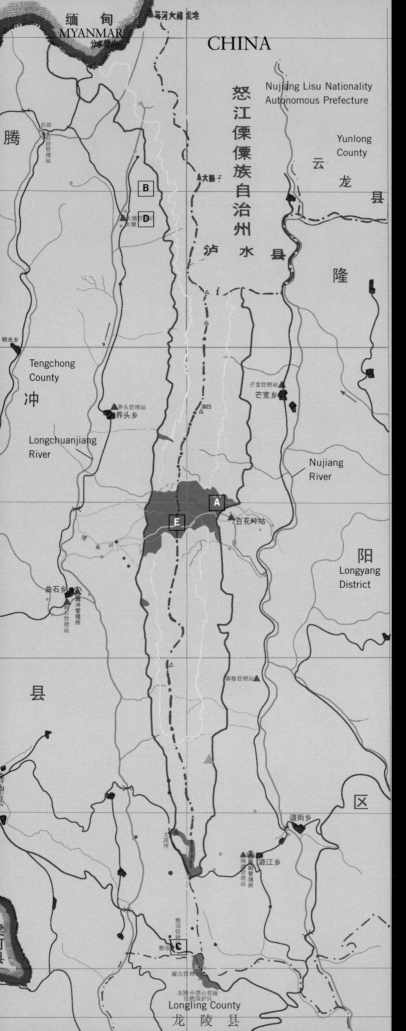

缅甸
MYANMAR

平河大垭塆地
分水管站

CHINA

腾

自治管理站

B
D
大塘管理站
大塘

大脑子

怒江傈僳族自治州

Nujiang Lisu Nationality
Autonomous Prefecture

Yunlong
County

云龙县

泸水县

冲

明光乡

Tengchong
County

界头管理站
界头乡

3823

芒宽管理站
芒宽乡

Longchuanjiang
River

Nujiang
River

A
E
百花岭站

阳

Longyang
District

曲石乡
高冲管理所
高冲管理站

赛格管理站

县

区

道街乡

大箐坪

潞江管理所
潞江乡

整顶管理所
整顶

C

崛元管理站

龙陵小黑山省级
自然保护区

Longling County

龙陵县

N

图 FIG. 3

Gaoligong Mountain National Nature Reserve
(Southern Section)

比例尺

├── 10 km ─── 20 km

图例 Legend

国界
International Boundary

县界
County Boundary

自然保护区界
Nature Reserve Boundary

核心区
Core Area

景区公路
Scenic Area Road

古道
Southern Silk Road

保护区内规划区域
Visitor-accessible
highlands

管理站
Baihualing
Management Station

管理所
Management Precinct

村寨
Village, Hamlet

公路 桥梁
Paved Road, Bridge

河流
River

县（市）政府驻地
County/Municipality
Seat

乡（镇）政府驻地
Village/Township Seat

图 FIG. 3A 植被的垂直变化主要跟云雾和干旱出现的频率和持续时间有关。Along the gradient of elevation, changes in vegetation are largely associated with frequency of cloud contact and relative lack of drought.

图 FIG. 3B 大塘附近高黎贡山脚下的村民可以积极参与保护区的管理，特别是低海拔地区的生态恢复活动。Villagers at the foot of the mountains in Datang can become active stewards of the reserve, especially of ecological restoration efforts in the lower elevations.

图 FIG. 3C 赧亢管理站，位于保腾公路旁保护区最南端的垭口处。The station at Nankang is at a low pass, along the highway that borders the southern limit of the National Nature Reserve.

图 FIG. 3D 保护区外稻田使用农用化学物质加速了两栖动物种群的下降；与当地村民合作开展一些试验活动可能有助于扭转两栖动物种群下降的趋势。Agricultural chemicals in rice paddies outside the reserve accelerate the decline in amphibian populations. Model projects with local farmers may help reverse the trend.

图 FIG. 3E 位于古丝绸之路上的南斋公房垭口（海拔3,100m左右）石头房，千百年来，此地一直是过往行人歇脚的地方。The stone shelter at Nan Zhaigongfang, the mountain pass at 3,100 m along the Southern Silk Road, has been a rest stop for travelers for hundreds of years.

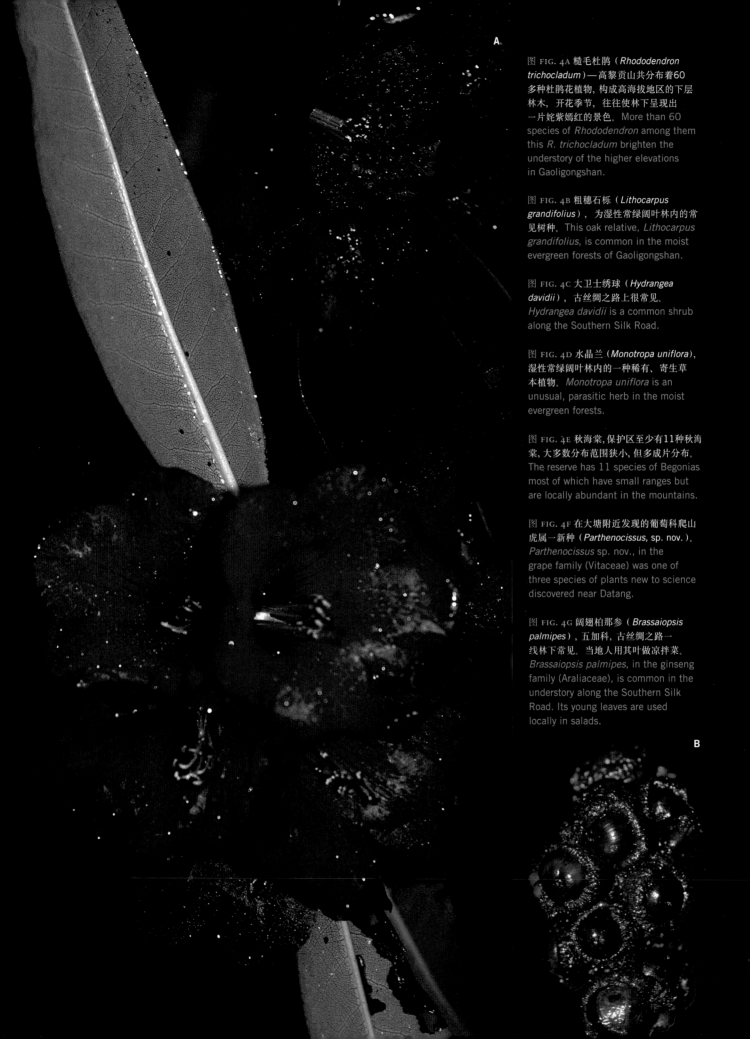

A

图 FIG. 4A 糙毛杜鹃（*Rhododendron trichocladum*）—高黎贡山共分布着60多种杜鹃花植物，构成高海拔地区的下层林木，开花季节，往往使林下呈现出一片姹紫嫣红的景色。More than 60 species of *Rhododendron* among them this *R. trichocladum* brighten the understory of the higher elevations in Gaoligongshan.

图 FIG. 4B 粗穗石栎（*Lithocarpus grandifolius*），为湿性常绿阔叶林内的常见树种。This oak relative, *Lithocarpus grandifolius*, is common in the moist evergreen forests of Gaoligongshan.

图 FIG. 4C 大卫士绣球（*Hydrangea davidii*），古丝绸之路上很常见。*Hydrangea davidii* is a common shrub along the Southern Silk Road.

图 FIG. 4D 水晶兰（*Monotropa uniflora*），湿性常绿阔叶林内的一种稀有、寄生草本植物。*Monotropa uniflora* is an unusual, parasitic herb in the moist evergreen forests.

图 FIG. 4E 秋海棠，保护区至少有11种秋海棠，大多数分布范围狭小，但多成片分布。The reserve has 11 species of Begonias most of which have small ranges but are locally abundant in the mountains.

图 FIG. 4F 在大塘附近发现的葡萄科爬山虎属一新种（*Parthenocissus*, sp. nov.）。*Parthenocissus* sp. nov., in the grape family (Vitaceae) was one of three species of plants new to science discovered near Datang.

图 FIG. 4G 阔翅柏那参（*Brassaiopsis palmipes*），五加科，古丝绸之路一线林下常见。当地人用其叶做凉拌菜。*Brassaiopsis palmipes*, in the ginseng family (Araliaceae), is common in the understory along the Southern Silk Road. Its young leaves are used locally in salads.

B

23

上一些最美丽、系统发育多样化的类群。虽然在高黎贡山只发现了一种蝾螈（图5a），但从低洼林地到4,000多米的高山却分布着至少6个科的蛙类和蟾蜍。

Gaoligongshan is a crucial reserve for amphibians. Its fauna is extremely rich, including some of the world's most beautiful and phylogenetically diverse taxa. Only one species of salamander (5a) is known from the mountains, but members of at least six families of frogs and toads occur from low-lying forests to headwater streams above 4,000 m elevation.

图 FIG. 5C 缅甸蟾蜍 *Bufo burmanus*
图 FIG. 5D 小角蟾 *Megophrys minor*
图 FIG. 5E 无指盘臭蛙 *Rana grahami*
图 FIG. 5F 滇蛙 *Rana pleuraden*

图 FIG. 6 保护区蛇类和蜥蜴主要分布在低、中海拔地带。所有的蜥蜴类和大多数的蛇类都是重要的捕食者。在有人类生活的地区，这些动物捕食昆虫和老鼠等有害生物，保护人类的家园和生产、生活。
The greatest species richness of snakes and lizards in the Nature Reserve is at low and middle elevations. All species of lizards and most of snakes are important predators. In human dominated landscapes, they eat insects and rodents that can be pests in homes or agricultural fields.

图 FIG. 6A 截趾虎 *Gehyra mutilata*
图 FIG. 6B 铜蜓蜥 *Sphenomorphus indicus*
图 FIG. 6C 颈槽蛇 *Rhabdophis nuchalis*
图 FIG. 6D 裸耳龙蜥 *Japalura dymondi*
图 FIG. 6E 蛇 *Ophites laoensis*

图 FIG. 7A 考察中发现的许多真菌都为新记录。该真菌为小皮伞属（*Marasmius* sp.），尚未定名。Many of the fungi we found are species new to science. This beautiful *Marasmius* sp. does not yet have a name.

图 FIG. 7B 小皮伞（*Marasmius purpureostiatus*），能分解枯枝落叶，高黎贡山可能是其分布的北限。*Marasmius purpureostiatus*, a litter decomposer, may reach its northern limit in southern Gaoligongshan.

图 FIG. 7C 杯冠瑚菌（*Clavicorona* sp.），其子实体与一个广布种的子实体非常类似，但是分子层次上的研究证明该种属中国南方特有种。The fruitbodies in this *Clavicorona* sp. are nearly identical to those of a broadly distributed species, but molecular studies show this southern Chinese species to be distinct.

图 FIG. 7D 鹅膏菌（*Amanita rubrovolvata*），属于温带、东亚分布种，与壳斗科植物形成重要的共生关系。*Amanita rubrovolvata* has a temperate, east Asian distribution; it forms a critical symbiosis with trees in the oak family.

图 FIG. 7E 中国橙色牛肝（*Boletus sinoaurantiacus*），一种最近才得到描述的种，仅分布在中国西南。*Boletus sinoaurantiacus* is a newly described species known only from southwest China.

图 FIG. 8A 在大理街头，妇女挑着各种水果向游客兜售。Abundant fruits—such as those carried by this woman in Dali—provide a colorful attraction for tourists.

图 FIG. 8B 当地村民用棕毛做蓑衣，防雨效果比较好。Fibers from palm fronds provide reliable, waterproof rain gear to local villagers.

图 FIG. 8C 芒市街头出售的各种香料和草药。Spices and herbs, used for medicinal and cooking purposes, abound in the outdoor markets of Mangshi.

图 FIG. 8D 在云南省山区，马是很好的运输工具。Yunnan horses are agile pack animals in the mountains.

图 FIG. 8E 有些村民还保留着制作传统绣花鞋的手艺。Some villagers still produce traditional embroidered shoes.

图 FIG. 8F 许多世纪以来，南丝绸之路一直是联结中国中原和中东的通道。古道的一段即穿过保护区。For centuries the Southern Silk Road connected central China to the Middle East. A portion of the ancient stone path traverses the higher elevations of the reserve.

图 FIG. 9A 入口处旅舍和游客接待中心候选地。此处靠近保护区边界，俯瞰怒江。 The proposed Gateway Lodge and Visitor Center site—just outside the reserve—overlooks the Nujiang River.

图 FIG. 9B 南方丝绸之路从村子里穿过。 The Southern Silk Road makes its way through local villages.

图 FIG. 9C 当地人至今还用这些具有独特图案的瓦来盖屋顶。 Clay roof tiles with unique designs still are used in the region.

图 FIG. 9D 传统的建筑风格和当地的建筑材料彰显出高黎贡山村寨的特色。 Traditional architecture and native building materials contribute to the character of Gaoligongshan villages.

图 FIG. 9E 当地村庄还保留着传统的房瓦制作工艺，并沿用这种古式瓦作房顶。 Historic tile roofs survive in local villages thanks to expert craftsmanship.

图 FIG. 9F 要保持传统的建筑风格，必须使用本地建筑材料，如花岗石。但是，这些材料多半需取自于生态敏感区域。 Indigenous building materials—such as this granite—are crucial for preserving local character, but must be quarried away from ecologically sensitive areas.

考察结果综述

生态特征

位于中国西南的高黎贡山山脉是一个极具特色的动态复合体 — 它既是一座巍峨高耸的地理屏障，又是各种动植物东西和南北迁移的主要通道。整个高黎贡山包括许多海拔在4,000m以上的山峰，是物种形成的温床。由于气候湿润、受冰川影响小，使得该地区特别是中、低海拔地区得以保存着许多原始物种。生态方面，高黎贡山具有其复杂性与独特性：它是地球上唯一保存有大片由湿润热带到温带森林过渡的森林类型的地区。在高黎贡山，有利于新物种的形成和防止古老物种灭绝的条件存在的时间长、地域广。此外，该地区也是许多古代商路和多元文化汇集的地方。

贡山国家级自然保护区位于高黎贡山山脉南段的中缅边境地区，面积为405,549公顷（图2，图3）。2002年6-7月我们在保护区及其周边地区开展了快速生物及社会考察活动，并对该地区开展生态旅游活动等需要的物质条件等进行了分析。在保护区周边地区居住着汉、傣、傈僳、回、白，苗、彝、壮、怒、阿昌、景颇、佤、德昂、纳西、独龙和藏16个民族，3万余人。

横穿高黎贡山的南丝绸之路（图8G, 9B）从公元前300年前开始就是中国通往印度、阿富汗等的交通要道。南丝绸之路从四川成都开始，贯穿云南全境到达保山市（古称"永昌"），翻过高黎贡山（图2），然后进入缅甸。由于所经之地气候温和，南丝绸之路全年通行无阻，使远在罗马的商贾、马帮能够到达中国。今天，在百花岭附近还可以看到许多这条古丝绸之路上的历史遗迹，如黄心树的石拱桥、直到1958年保腾公路修通以前一直在发挥作用的旧街子等。

高黎贡山地区也是二次大战期间的主要战场之一。当时盟军士兵和中国劳工在极短的时间内修通了滇缅公路，使战略物质能够从当时处于英军控制之下的缅甸运送到中国。中美军队在这一带击退了侵入中国的日本军队，位于古丝绸之路上的南斋公房垭口就是当年的许多战场之一。

虽然这条石头铺就的道路（图 8F, 9B）几千年来一直见证着繁忙的马帮、商队和旅行者，但是，在海拔2,000m以上的地区，古道两旁的植被特征却没有发生明显的改变。一个例外可能就是南斋公房附近(图3E)。由于长期牲口啃食、践踏和过往行人拾柴生火等的干扰，海拔虽然只有3,200m，离正常的树线还有1,000多米，但此处附近的主要植被却是低矮的灌木丛。

在保护区海拔2,000m以下的地区，当地村民多年来一直在里面放牧、砍树，森林利用程度较高（图 2B, 2C, 3D）。即使那些未被砍伐的有林地带，森林也高度退化。但是，就是在这些退化的森林里，生物多样性也非常丰富。这些低海拔地区物种丰富。在考察中我们就发现了一些具有较高研究价值、分布范围狭窄的动植物物种，包括至少3种尚未见报道的植物。

高黎贡山的岩层主要是高度侵蚀的花岗岩。东坡的植被类型带有典型的酸性花岗基岩特征。而在西坡，由于火山灰的作用，植被群落发育的地球化学环境与东坡不同，酸性较弱。考察中发现，东西坡在动植物和真菌等的种类组成上都存在明显的差异。物种随海拔高度的变化也极大地增加了保护区物种的丰富程度。

高黎贡山的冬季（每年11月到笠年4月）一般比较凉爽、干燥，而夏季（5-10月）则温暖多雨。山坡上主要分布着季风常绿阔叶林，其间镶嵌着一些针叶林或混生着一些针叶树木，大大增加了保护区内生境类型的丰富程度。在开花季节，丰富的杜鹃花科植物构成高黎贡山特别是在酸性土壤占优势的东坡下层林木的一道特别靓丽的风景(图1,4A)。在海拔较高的地方，竹类构成下层林木的优势种类 (图2F)，在海拔稍低的地方，竹子则侵入到受到干扰的区域。

高黎贡山东西坡有无数的山间小溪和短小的河流自保护区流出，汇入怒江或龙江。这些小溪和河流在保护区形成道道沟壑和众多飞瀑流泉（图2A）。这些快速流动的河流（图2E）本身也是许多动物包括一些仅见于高黎贡山的鸟类和两栖动物的关键生境。这些沟壑也分布着许多截然不同的植物群落。大多数分布在谷底的植物都依赖鸟类来传播种子，而山脊上则主要分布着依靠哺乳动物或风媒传播的植物。

虽然我们在保护区的三个地点进行了短暂的抽样调查，却已经发现了几个新种，足见该地区物种的丰富程度以及在该地区发现新的物种的潜力。在文化调查中（以前在该区域尚未开展过此类调查），我们进一步认识了当地村寨丰富的文化资源，了解了当地开展生态旅游活动的潜力，评估了当地村寨在开展旅游活动方面需要改善的领域。我们还利用考察的机会为当地设计了一个生态旅舍和游客接待中心。这个初步设计方案充分利用目前的百花岭管理所的现有设施，考虑了与当地的环境的和谐，也考虑了游客的需求。

在报告的下列几节中，我们将对考察结果的要点进行总结，并提出我们对在当地开展保护行动和生态旅游活动的一些建议。

高黎贡山保护工作回顾

二十世纪早期，两位世界知名的博物学家，Joseph Rock（当时在美国国家地理学会工作）和 George Forrest（其时于爱丁堡植物园工作）在考察高黎贡山 包括百花岭地区的时候都有很多重要发现。

1983年，云南省批准成立高黎贡山省级自然保护区，1986年被升级为国家级自然保护区。2000年保护区被联合国教科文组织接纳为人与生物圈保护区网络成员。同年，保护区进行了比较大的拓展，目前正在酝酿进一步扩大保护区的范围。保护区发展的一些主要事件包括:

1983年: 云南省政府批准成立高黎贡山自然保护区，保护区横跨保山地区和怒江傈僳族怒族自治州。

1986年: 国务院批准成立高黎贡山国家级自然保护区，并对保护区的边界进行了调整。保护区总面积123,900公顷，南北长135公里，东西宽平均9公里。

1994年: 国务院批准允许在保护区的一小部分（约8,550公顷，即21,119英亩）开展旅游活动。开展旅游活动的区域主要限定在三处: 北段的片马（3,649 公顷），西坡的大树杜鹃区域（约270公顷）和百花岭周围（4,361公顷）。百花岭被确定为第一个开展旅游活动的区域。

2000年: 国务院大幅度拓展了保护区的范围，使保护区面积扩大了三倍，达到 405,549公顷，主要将原来高黎贡山北段的一个省级保护区与原高黎贡山保护区合并。

2002 年: 人们提议进一步扩大保护区面积，将从怒江到龙江的广大领域都纳入保护计划（图3）。

百花岭村: 资源与能力

Victoria C. Drake Anne P. Underhill 朱明育 郝光明 杨建美

在当地向导和翻译的帮助下，我们开展了为期13天的社会调查。百花岭村包括8个自然村: 汉龙、大鱼塘上社、大鱼塘下社、帮外-古兴寨、桃园、老门寨-百花岭-麻栎山、芒岗和芒荒（见附录6 表1）。每个自然村都有一个小组长。在每个自然村里面，我们挑选了小组长家和一户普通村民进行访谈，目的是了解村民开展旅游活动并从中受益的能力。访谈中讨论的主要问题及访问结果总结见附录6。

在历史的长河中，各个自然村都发生了许多变迁。由于人口的增长，有些寨子分成了几个小的寨子；也有些小的寨子由于不断扩展而合成了一个大的寨子。百花岭行政村有大约450户、2,100名村民。每个自然村都有水田和旱地。百花岭村村委会主任也是该村的党支部书记，村委会办公室在百花岭自然村。

民族文化——今日百花岭地区的主要民族有汉族、傣族、回族、白族、傈僳族和彝族。虽然这些民族都保持着各自的民族文化，但在历史的进程中，文化融合是难以避免的，而且随着人们对民族间通婚的逐渐认可，在过去二十年里，这种融合的过程加快了。在考察中我们发现大家对自己的民族身份持非常灵活的方式。由于在很多家庭里父母可能来自不同的民族，小孩子可以随父亲的民族，也可以随母亲的民族。由于婚姻关系往往使几世同堂的家庭模式发生改变，所以各户的居住地的情况也相差悬殊。如果女方家里没有儿子，则可以采取招亲的方式使男到女家。

虽然有些人会说傈僳话或彝族话，但是，在很多场合大多数人都讲汉语，而不是他们自己的民族语言。如果有看得见的好处，有些人也会学习他们自己的民族语言。例如，有些人为了更好地同傈僳族同胞做生意便学习傈僳语。传教士为傈僳族同胞创立了文字，芒岗附近的教堂使用的就是用傈僳语写的圣经。由于需要从事艰苦的劳动，如在崎岖、多山的高黎贡山谋生，许多民族已经放弃了本民族的传统服饰、烹饪习惯和手工艺品的制作。大多数人家都过汉族的春节，但同时也过彝族的传统节日，如火把节。不过，深入的研究将会发现各个民族在习惯与信念方面的差异。

自然资源与挑战——每个村子都有不同的社会条件，人们都面临着谋生的各种挑战（附录6表2）。据我们估计，百花岭村的年户均收入在300到20,000元之间，平均约2,000元。所有农户都感到农业生产对他们压力大。接受我们访谈的村民都说他们希望能够有更多的钱来提高农业活动的科技含量和购买化肥。住在平缓地区的村子的水田比较多。绝大多数农户种植水稻，作为自家的口粮。现金收入大部分（85%以上）来源于种植和出售甘蔗。农户的收入与附近的糖厂的经营状况息息相关，糖厂为农户提供蔗种，同时收购农户的甘蔗。当地交通不便，道路比较陡峭而且下雨时通行非常困难，村子里汽车、轿车和其它交通工具很少。

所有农户都面临烧柴短缺的问题，几乎所有家庭都采集枯木和灌木作为烧柴。迫切需要开发如沼汽等替代能源。限制薪柴采集的一个因素是土地所有（使用）权。汉龙村距离保护区管理站和业已存在的旅游线路最近，也最需要关注保护区森林的保护。有些农户采集菌类和草药补贴经济收入（图 8A, 8C），而有些村民则靠采集和出售野生蜂蜜获取现金。百花岭 村90%以上的农户使用了节柴灶，使能源消耗减少了60%。而在麻栎山则有一些农户还在继续使用不节能的"老虎灶".该地区开展旅游活动肯定会对薪柴资源带来更大的压力。

虽然村民反映怒江的水比前几年要清澈了，但是农药、化肥和鼠药等的过度施用肯定还在继续影响水的质量。

每个村民在应付基本的开销如购买化肥、农药、小孩上学、求医看病、处理红白喜事等方面都显得捉襟见肘。各个村子的人们都有自己应付家庭收入不足的方法（附录6表3）。至于哪个村子经济条件最差，各人的感觉不一。不过老蒙寨和麻栎山有更多的村民表示他们最需要在农业上得到帮助。有些村民已经认识到作物多样化、环境保护和经济收入之间的关系。部分农户的实践证明芒果、荔枝和桔子等果木在当地的产量能够达到比较高的水平。

文化和历史资源——各个自然村都存在不同的现代文化资源和古代历史遗迹（附录6表4）。 高黎贡山国家级自然保护区人类居住的历史可以追溯到两千五百年以前的哀牢人。哀牢古国以现在的保山为中心，兴盛于公元前500-100年左右。目前该地区尚存的古代先民留下的物证建于明代（1368-1644年）。一个特别引人注目的地点就是在汉龙村附近的古丝绸之路上的一处残垣断壁。另外就是在现在的观音寺附近的一些断碑上的明代碑文。根据当地人讲，这些碑是从原来的观音寺搬到这些地方来的。今后应该开展进一步的调查，对明代时期关于高黎贡山的记载进行研究。汉龙村村民吴先生建立的抗日战争博物馆对游客也具有一定的吸引力，不过需要进一步完善，因为博物馆中有一些还没有拆掉引线的二战时期留下的炸弹。有些村民还保留着许多传统手工技术，制作精美的绣花鞋（图8E）、围裙、篮子、瓜瓢、柳条框、马鬃雨披和马鞍等。

社会资源——在百花岭村已经有一些团体在开展发展当地经济、促进自然保护的活动（见附录6表5）。在MacArthur基金会的资助下，当地在1995年成立了中国第一个农民生物多样性保护协会。协会成立的目的是为了更好地保护环境。许多农户都希望被选为示范户，以便加入到经济发展和自然保护的过程中。百花岭村民还自发地成立了许多颇有独创性的组织，如大鱼塘的旅游协会等。

大多数村子的村民都非常重视下一代的教育（附录6 表6）。由于交不起学费，村子里很少有人能够读完高中或上大学。为了攒钱送子女上学，许多家庭不惜出售自己的家产或与别人换工。学生毕业后在村子以外的地方找工作比较困难，除非政府安排。但是，还是有些人看到了教育、经济发展和自然保护之间的关系。

村民关于开展旅游活动的看法——有些村民也意识到了旅游发展、自然保护和经济发展之间的联系（附录6表7）。不过，很少村民同旅游者打过交道，也不了解旅游活动会对他们产生什么影响。对开展旅游活动比较热心的家庭也都缺乏启动资金。村民还希望保持当地独特的自然景观。在生产传统工艺品和食品的时候还必须考虑环境保护。人们欢迎举办培训班，帮助人们认识旅游活动以及发展旅游与自然保护之间的关系。要帮助人们正确认识旅游活动并从旅游活动中获益，教育是关键。村民希望在当地的发展中有一定的发言权，他们对旅游设施的地点与管理和随着旅游的开展如何继续对保护区进行保护等话题也比较关心。

植被与植物区系

文军 Robin Foster 覃家理 孟世良 邓莉兰 施晓春

高黎贡山南段的植被类型异常复杂。除了海拔高度以外，历史因素特别是大的人为干扰直接影响了植物种类构成。此外，土壤等也是影响植物组成的一个重要因素，

例如，在中海拔地区的矮林就与土壤有直接的关系。我们考察的三个地点即百花岭、大塘和赧亢（图3）的植物群落有明显差别，植被构成和海拔高度、底质（岩层和土壤）、地形和人为干扰（类型、强度和时间）等有直接的联系。

从低海拔到高海拔、从东坡到西坡，植物组成变化显著。雄起的山峰成了物种东西向扩散的屏障，而在山脉南端海拔较低的地方，物种又有可能相互迁移。地质差异也是高黎贡山东西坡植物组成出现差异的原因之一：高黎贡山东坡为酸性花岗岩，而在西坡，在花岗岩的上面沉积有大量的火山灰，土壤的酸性较弱。例如，在土壤酸性较强的东坡，杜鹃花属（Rhododendron spp.）（图4A）植物的种类就比西坡多，种群也西坡大；而槭属（Acer spp.）植物在东西坡则分布着不同的种。

此外，在同一坡面，由于土壤底质呈小块镶嵌式分布，也增加了高黎贡山的总体植被多样性，不同底质的土壤上分布着不同的物种。例如，分布在山脊上的石英砂岩与粘土上的杜鹃矮林，虽然在同一片林子里的物种比较单纯，但是不同位置的杜鹃林的物种组成有很大的差别。底质对物种构成的影响在从百花岭管理站到瀑布（图2A）的路上表现得非常明显：最靠近瀑布的山脊上的植物成分与距瀑布稍远的山脊上的植物构成相比，只有10%是相同的。

随着海拔的升高，植被的变化大多与云雾笼罩的情况和水分的充足情况有关（图 3A）。在百花岭以上地区，有两处比较明显的植被过渡带：一处在海拔2,000m左右，另一处在海拔2,600m和2,700m之间。海拔2,000m左右刚好是夏季长期被云雾遮盖的地方，而海拔2,600m至2,700m之间则是冬季云雾经常到达的海拔高度。由于山脊和沟谷在水分上的差异，即使在同一海拔高度上，植物群落也有很大的变化。在海拔最高的地方，植被类型主要受温度、风和雷电等因素的影响。在突出的山脊地带的针叶树和上层树木上可以看到明显的、遭受频繁雷击的迹象。

海拔较低的山坡上的物种要比高海拔地区的物种丰富得多。有趣的是，此次考察中发现的一些不寻常的物种或该地区的新记录物种也是在低海拔地区采集到的。海拔较低的地区已经在保护区边界以外，受到周边地区人口的严重干扰，而且这种人类扰动还在不断加剧（图2C，见威胁部分）。

在高黎贡山, 滑坡似乎不太频繁, 规模也比较小, 而在其它许多山地森林中, 频繁而大规模的山体滑坡经常是影响森林格局的主要自然干扰因子。虽然火灾也比较少见, 但是, 在保护区确实可以看到火烧影响森林动态的情况, 特别是山脊和矮林地带。动物对种子的传播对植物组成有很大的影响。海拔较高的山脊和山坡上主要是靠动物传播种子的植物, 而在海拔较低的山坡和谷底地带, 则以靠鸟类传播种子的植物占优势。植物种类的丰富程度及其繁殖的季节性对动物区系的构成有很大影响, 反之亦然。

除了杜鹃矮林以外, 其它物种也构成一些地方的优势物种。在中海拔沟谷地带, 我们发现大片茂密的同龄马蓝属 (*Strobilanthes*, 爵床科) 居群, 本属植物一般在达到10-20年龄后开花然后枯死。在保护区内还有许多茂密的竹林, 其生态学特征与马蓝属植物类似, 在开花结实的年份能够侵入并定居在滑坡形成的裸露地段。

植物区系丰富度与植物种类组成——高黎贡山的植物区系的物种丰富度与特有度都比较高 (图4)。植物考察小组估计在整个高黎贡山山脉可能分布有5,000多种高等植物, 其中有4,500种左右已经被记录和描述 (210科, 1,086属, 4,303种或变种, 薛纪如等,1995; 李恒等, 2000)。在考察中至少发现了三个新种, 其中2种为五加科 (Araliaceae) 植物, 1种为葡萄科 (Vitaceae) 植物 (见*Parthenocissus*爬山虎属一新种, 图 4F), 都是在大塘发现的。在全部物种中, 约有10% (434种) 为高黎贡山特有。在为期9天的短暂考察中, 我们一共观察到了1,000多种高等植物 (包括6个新记录), 采集到300种植物标本, 并对250多种植物拍了照片。绝大部分尚未被描述或记录的物种主要分布在一些研究得比较少的植物类群, 如蕨类植物、兰科 (Orchidaceae)、蔷薇科 (Rosaceae)、杜鹃花科 (Ericaceae)、萝摩科 (Asclepiadaceae)、葡萄科 (Vitaceae)、荨麻科 (Urticaceae)、唇形科 (Labiatae)、苦苣苔科 (Gesneriaceae) 和山茶科(Theaceae) 等植物。保护区的蕨类植物非常丰富 (附录 2), 而且迄今为止尚未被系统研究过。许多分布在中山常绿林内。在考察中我们一共采集到了30科和52个属的蕨类植物。其中5科(木贼科 (Equisetaceae)、铁线蕨科 (Adiantaceae)、槲蕨科(Drynariaceae)、剑蕨科 (Loxogrammaceae) 和 满江红科 (Azollaceae)) 和8属(问荆(*Equisetum*)、毛轴碎米蕨 (*Cheilanthus*)、假棕

毛粉背蕨 (*Aleuritopteris*)、鞭叶铁线蕨 (*Adiantum*)、滇南狗脊蕨 (*Woodwardia*), 槲蕨 (*Drynaria*), 中华剑蕨 (*Loxogramme*), 和 满江红 (*Azolla*))为高黎贡山新记录。这些新的记录都是在大塘采集到的。双扇蕨科 (Pteridaceae)、鳞毛蕨科 (Dryopteridaceae) 和水龙骨科 (Polypodiaceae) 等科的蕨类物种特别丰富, 我们相信在这些科里面肯定还有大量的新种等待人们去发现。

考察的三个地点的情况——(1) 百花岭: 在百花岭的考察几乎包括了东坡从怒江边沿古丝绸之路一直到南斋公房的一条完整的植物群落的垂直样带 (图2, 3)。在从百花岭海拔1,525m到南斋公房海拔3,100m之间的这条样带上, 至少有三个与海拔高度或地形地貌有关的主要植被带。由于底质不同, 每个植被带都有至少两个以上的差异明显的群落类型。

在距离管理站不远的温泉附近为季风常绿阔叶林。在黄竹河 (古丝绸之路上的一个小站, 考察队在此宿营) 附近海拔2,000m至2,800m的地带分布着中山湿性常绿阔叶林, 林类附生植物, 特别是兰科 (Orchidaceae), 五加科 (Araliaceae), 及水龙骨科 (Polypodiaceae) 异常丰富。这一带壳斗科 (Fagaceae)、樟科 (Lauraceae)、木兰科 (Magnoliaceae) 和杜鹃花科 (Ericaceae) 的植物种类繁多。主要物种包括多变石栎 (*Lithocarpus variolosus*)、野八角 (*Illicium simonsii*)、核桃 (*Juglans regia*)、栲 (*Castanopsis* sp.)、青冈 (*Cyclobalanopsis lamellosa*)、杜英 (*Elaeocarpus* spp.)、红花木莲 (*Manglietia insignis*)、槭 (*Acer* sp.)、多花含笑 (*Michelia floribunda*)、西南锈球 (*Hydrangea* sp.)、箭竹 (*Fargesia edulis*)、假柄掌叶树 (*Brassaiopsis palmipes*)、浅裂掌叶树 (*Brassaiopsis hainla*)、常青木(*Merrilliopanax listeri*)、山矾 (*Symplocos* spp.)、越桔 (*Vaccinium* spp.)、蕨类 (*Pteris* spp.)、以及杜鹃花属 (*Rhododendron*) 的几种植物。在考察队宿营的南斋公房垭口 (图3E) 附近, 海拔2,800m至3,200m的山坡上, 植被主要为灌木林地, 主要植物种类有蔷薇 (*Rosa* sp.)、悬钩子 (*Rubus* sp.)、花楸 (*Sorbus* sp.)、龙胆 (*Gentiana* sp.)、马先蒿 (*Pedicularis* spp.)、杜鹃 (*Rhododendron* spp.)、石栎 (*Lithocarpus* spp.)、箭竹 (*Fargesia* spp.)、多齿十大功劳 (*Mahonia polyodonta*)、瑞丽鹅掌柴 (*Schefflera shweliensis*)、滇瑞香 (*Daphne* sp.)、尼泊尔酸模 (*Rumex nepalensis*)、及驴蹄草 (*Caltha palustres*) 等。

(2) 大塘：大塘在高黎贡山的西坡，与百花岭一线相比，研究得比较少。尽管大塘附近的植被遭受了严重的人为干扰（伐木、种植烤烟、放牧和采集薪材），但是，大塘的物种仍然十分丰富。考察队在此发现了不少具有特殊意义的植物，包括3个新种。在我们记录到的一些分布范围狭窄的物种中有大树杜鹃（*Rhododendron protistum* var. *giganteum*）、秃杉（*Taiwania flousiana*）、树蕨（*Alsophila spinulosa*）、水青树（*Tetracentron sinense*）、多变三七（*Panax variabilis*）和总序五叶参（*Aralia lihengiana*）等濒危植物。考察中记录到的优势种类有石栎（*Lithocarpus* spp.）、粗梗稠李（*Prunus nepalensis*）、青榨槭（*Acer davidii*）、薄叶虎皮楠（*Daphniophyllum chartaceum*）、美丽马醉木（*Pieris formosa*）、吴茱萸（*Evodia* sp.）、香叶树（*Lindera communis*）、藏滇猫儿屎（*Decaisnea insignis*）、水青树（*Tetracentron sinense*）、红花木莲（*Manglietia insignis*）、含笑（*Michelia* sp.）、滇结香（*Edgeworthia gardneri*）、多花酸藤子（*Embelia floribunda*）、爬山虎（*Parthenocissus* spp.）、崖爬藤（*Tetrastigma* spp.）、小花香槐（*Cladrastis sinensis*）、厚皮香（*Ternstroemia gymnantheca*）、白木通（*Akebia trifoliata*）、峨眉蔷薇（*Rosa omeiensis*）、油葫芦（*Pyrularia edulis*）、鸢尾（*Iris tectorum*）以及 杜英（*Elaeocarpus*）属的几个种。

(3) 赧亢：赧亢位于高黎贡山国家级自然保护区南端的垭口（图3）。由于该处海拔相对较低，约 2,150 m，所以是高黎贡山东西坡植物交流的重要生物走廊。考察中我们发现，赧亢的植物与东坡的百花岭和西坡的大塘都有许多相似之处。此处常年多雾多雨，森林为中山湿性常绿阔叶林，兰科和蕨类等附生植物很丰富。优势森林物种包括多变石栎（*Lithocarpus variolosus*）、印度木荷（*Schima khasiana*）、小叶楠（*Phoebe* sp.）、栲（*Castanopsis lamellosa*）、含笑（*Michelia velutina*）、尾叶樟（*Cinnamomum caudiferum*）、马缨花（*Rhododendron delavayi*），大白花杜鹃（*Rhododendron decorum*）、高鹅掌柴（*Schefflera elata*）、红花木莲（*Manglietia insignis*）、瑞丽山龙眼（*Helicia shweliensis*）、大花八角（*Illicium macranthum*）和 紫花鹿药（*Maianthemum purpurem*）等。

大型真菌

Gregory M. Mueller　杨斌　葛尚义

高黎贡山大型真菌种类异常丰富。虽然这方面的研究十分有限，但到目前为止已经记录到了300种左右。在9天的考察中，我们只对有限海拔高度上有限的几种植被类型进行了调查，却已经记录到了200多种真菌（对其中148种制作了标本并拍摄了彩色照片，附录1，图7）。在我们记录到的种类中，只有22中被列入了先前的高黎贡山真菌名录（共记132种）。虽然无法准确估计保护区内的大型真菌（子囊菌、伞菌、檐状菌，珊瑚菌等）的种类，但是根据当地物种多样性程度、植被类型和海拔范围等，我们估计保护区内的大型真菌在1,500至2,000种左右。

不同区系的大型真菌在高黎贡山交融存在，使得该地区在大型真菌的保护和进一步研究方面具有国际重要性。调查中发现*Dictyopanus*（网孔菌属，一种热带分布的属）的两个种与许多温带种类（云南奥德菇（*Oudmansiella yunnansis*），黑乳菇（*Lactarius lignyotus*）和漆腊蘑（*Laccaria bicolor*））等，可能已经是这些物种的最南端的分布）共同生长，还发现温带亚洲种（鹅膏菌（*Amanita rubrovolvata*）（图7D）和 疣柄牛肝菌（*Leccinum virens*）与温带欧洲种 如奥德菇（*Oudmansiella muscida*）生活在一起。我们还记录到了一些应该是属于北美东部和亚洲东部间断分布的真菌种类，如干菌（*Xerula furfuracea*）。考察中还采集到了一些中国特有种类，如仅分布在中国南方的大而色彩鲜艳的中国橙色牛肝（*Boletus sinoaurantiacus*）（图7E）。

在采集到的148种标本中，只有2个种在三个考察地点都发现。很难说这种低重复率是由于考察地点生境的特殊性还是由于抽样调查的范围有限造成的。不过，各考察地点的物种有明显的差异，这一点是确定无疑的。真菌的分布与植被类型、土壤类型和水分供给等有密切的联系。高黎贡山地质地貌复杂、植被类型丰富，应该具有许多独具特色的大型真菌群落。在考察期间，我们没有采取随机调查的方式，而是由当地向导带领我们到他们认为菌类比较多的地方。

大型真菌的子实体生长有很强的季节性，在雨季的不同季节进行考察会有不同的调查结果。另外，大型真菌生长的年变化也比较明显，所以，即使在同一地点，不同的年份也会有不同的记录。所以，我们建议开展更加广泛的调查活动。虽然此次调查由于时间有限我们没有进行定量记录（每个调查点可能需要至少3-5天），但是，我们还是建议以后开展此项活动。每年至少两次，分别在雨季开始和结束时进行。通过几年连续的调查、收集数据，就能够对高黎贡山保护区内各区域以及保护区内外的真菌丰富程度和物种多样性进行比较。

真菌对森林和草地群落至关重要。保存其种类多样性和种群数量对于维护动植物群落具有关键的意义。大型真菌在促进养分循环与吸收、调节水分、保持植物以及其它土壤有机物之间的相互作用方面起着十分重要的作用。许多大型真菌与高黎贡山的优势树种如壳斗科、松科等具有共生关系。这些外生菌根群落对于树木和真菌的生长都是必不可少的。树木为真菌生长提供糖类，而真菌则供给树木矿物质和水分，保护树根免受病原体的侵袭。真菌还能够使树木的地下部分 — 根系相互联系起来，使它们共享碳水化合物和矿物质并形成一个动态的相互作用的群落。有些真菌是重要的分解者：大型真菌以及一些细菌是唯一能够分解纤维素和木素（构成植物体的两类基本物质）的生物。这些真菌是植物群落的初级循环者。也有一些真菌是树木的病原体，它们的活动促进了森林年龄结构的多样性。

只要管理和利用得当，大型真菌可以成为高黎贡山地区一项重要的、环境友好的替代经济来源。这些大型真菌为当地人提供了食品和药物，也是当地重要的经济来源。从生态旅游的角度，除了作为游客的美味佳肴外（对于外国游客来说无疑是一种异国风味），这些绚丽多彩的大型真菌和蘑菇（图 7）一定会大大丰富生态旅游者的旅游体验。

真菌可以在高黎贡山附近的社区和森林群落之间起到很好的联结作用。在美国西北部和加拿大西南开展的研究表明森林中的非木材产品（主要是食用或药用菌类、蕨类和森林蔬菜等）的经济价值完全可以超过森林提供的木材产品的价值。真菌一般大量生长于高质量的、管理比较好的天然植被群落里。作为一种有开发价值的可更新资源，真菌能够给周边地区的群众带来直接的好处。

两栖、爬行类动物

H. Bradley Shaffer   张　宇   王天灿

在不到7天的调查中，我们一共发现了30种两栖爬行类动物（附录3），包括7种蛇类、4种蜥蜴、1种蝾螈和18种蛙类（实际上可能在15-21种之间，要等进一步鉴定结果；图5，图6）。据已有文献（薛纪如等，1995）记载，高黎贡山地区共有72种两栖爬行动物。但是，在短暂的考察中，大家就已经发现了5种（约占总数的15%）原来的名录没有记录的物种，分别是滑鼠蛇（*Pytas mucosas*）、颈槽蛇（*Rhabdophis nuchalis*）（图6C）、红脖颈槽蛇（*Rhabdophis subminiatus*）、背条跳树蛙（*Chrixalus doriae*）和刘氏小岩蛙（*Micrixalus liui*）。据此推算，高黎贡山国家级自然保护区内肯定还有相当数量的物种没有得到记录。根据高黎贡山、云南以及云南省西南山区已经记录到的物种数量推测，我们估计高黎贡山保护区应该有大约60种爬行动物（以蛇类为主）和60种两栖类动物（主要为蛙类）。即使与物种丰富的亚马逊热带地区相比，这里的物种多样性也是比较高的。由于其活动隐秘，我们相信尚未发现的蛇类和蜥蜴类一定还比较多。对于保护区内蛇类多样性，我们可能还只是触及了皮毛。我们还怀疑原来人们认为分布广泛的一些蛙类中可能包含很多分布狭窄的表型相似种或隐秘种。

考察中，大家在夜间寻着鸣叫声对溪流、水塘和稻田等地进行调查。一般情况下，考察小组的三个成员在同一考察地点分别沿着不同的方向进行调查，确定发出鸣叫声的所有物种的位置。此外，还利用手电和头灯等对那些没有发出鸣叫声的动物进行了考察。在白天，考察队员沿着小路或路边调查，翻开倒木和石头，在灌木和草丛中寻找蜥蜴和蛇等。大家还沿着小溪和水塘等行走，查看蛙类或蟾蜍的成体或蝌蚪等。

高黎贡山汇聚了一些重要的中山亚热带种类（1,500-2,400 m）和高山种类（海拔 2,700 m以上）。高山两栖类物种分布的地理范围十分有限，尤以锄足蟾科（Pelobatidae）的分布最为狭窄。例如，贡山齿突蟾（*Scutiger gongshanenis*）就只分布在高黎贡山海拔2,500 m以上的地方的清澈、快速流动的溪流附近；沙巴拟髭蟾（*Leptobrachium chapaense*）、贡山树蛙（*Rhacophorus gongshanensis*）和云南龙蜥（*Japalura yunnanensis*）等也仅分布于高黎贡山地区。但是，由于考察时间有限，我们没有发现这些物种。我们认为，这些分布在中、高海拔地区被人们处理为一个物种的东西很可能是几个不同的物种，进一步的研究可能还会发现许多只分布于高黎贡山的种类。

在海拔 2,700m 左右，两栖爬行类群落构成一下子从低地物种转为高山物种类型。在 2,500m 甚至 2,700m 的高度，考察队员都还听到小角蟾 (Megophrys minor) (图 5D) 的鸣叫，再往上就没有了。到了海拔 3,100m 左右，在低地一般为小角蟾占据的溪流边，考察人员听到了贡山齿突蟾 (Scutiger gongshanenis) 的清晰鸣叫声。在此海拔高度，我们还听到了另外一种在低海拔地区未曾听到过的蛙类的鸣叫声，但是由于没有捕捉到实物标本，我们无法判断是哪个种。海拔 3,100m 左右的气候和森林结构都不适合低海拔物种如黑线乌梢蛇 (Zaocys nigromarginatus) 和滑鼠蛇 (Ptyas mucosas) 的生存。

虽然野外考察时间短暂，但是调查结果还是足以说明一些问题。首先，高黎贡山本身可能就是一些两栖类动物扩散的主要生物地理屏障。除了采集两栖类标本以外，我们还根据其鸣叫声对其种群数量进行了估计。例如，在百花岭一带海拔 1,400m 到 2,500m 之间的范围类，几乎所有的宽度在 1m 以内的水沟都能够听到小角蟾 (Megophrys minor) (图 5D) 的鸣叫。然而，在西坡的大塘类似的河流水沟边，我们从未听到过它的鸣叫声。该物种即使在西坡有分布，数量一定也比东坡少得多。所以我们估计，该物种在西坡根本没有分布。与此类似，在大塘和赧亢极其丰富的滇蛙 (Rana pleuraden) (图 5F) 在百花岭却没有。最后，红瘰蝾螈 (Tylototriton verrucosus) (图 5A) 在百花岭很多，即使在受到严重扰动的村子附近的水沟边也很常见；在赧亢虽有分布，但数量不多；而在大塘较少受到干扰的生境里都没有见到。由于该物种在保存完好和受到严重干扰的生境中都有分布，所以我们相信，如果真有分布的话，我们应该能够发现它们。

我们还认为高峻的山脉也是爬行动物扩大分布区的生物地理屏障。不过，在考察中发现的爬行动物没有两栖类多，因此，下此结论可能为时尚早。调查发现爬行动物与两栖类一样有呈镶嵌式块状分布的特点，但是，由于发现的数量较少，我们还不能肯定此次调查的结果能反映实际情况。

显然，要认识这些群落特别是爬行动物变化的范围和程度需要更加广泛的研究和调查。实际上，对蜥蜴和蛇类进行抽样调查的唯一有效的方法就是用带网兜的漂流栏网。这种工具安全简单，操作方便，也比较容易清点捕捉到的物种和数量。不过，这需要训练有素的人员（保护区管理人员或村民）每天检查，清点捕捉到的种类和数量然后放生。每检查一处可能需要大约一两个小时的工夫。不过，在人手不够的时候，可以很容易地取消捕捉装置，让动物自由通过。一旦有空，又可以开始监测活动。

考察队的许多专家都发现高黎贡山周围地区特别是低洼的农地里的两栖类的物种数量和种群数量都在减少。这与世界范围类两栖类的减少的趋势是一致的。造成两栖类动物种类和数量急剧减少的主要原因之一就是农药和除草剂的使用。农药和除草剂杀灭了两栖类的成体、蝌蚪和卵。高黎贡山保护区内有些湿地水塘、水池和溪流没有受到这些农药的影响，同保护区为的农田区域相比，一些常见的两栖类动物就要丰富的多。例如在大塘的管理站附近的农田和李家窝棚（距离管理站 5 公里左右）宿营点附近以及在赧亢我们都发现了类似的物种（图 3D）。但是，在农田里的物种数量少，每种动物被观察到的数量也很少。在 2 个小时的夜间观察中，每种动物只发现了 1–2 只个体。相比之下，在李家窝棚附近的水塘里，在同样长的时间类，我们观察到的两栖类动物的数量是在农田里观察到的 2–5 倍。在赧亢未受到干扰的植物园内，考察队员观察到的数量是在大塘管理站附近的农田里观察到的数量的 5–10 倍之多。由于免受农药之灾，高黎贡山国家级自然保护区为蛙类动物提供了极其重要的避难所。

对于生活在中、高海拔的两栖爬行动物来说，高黎贡山是一个很好的庇护所。在低海拔地带，我们相信大多数物种都存在，但是由于人类频繁活动，数量已经减少。减少人类活动的影响，或者改变人类活动的方式，是能够使低海拔地区的物种得以恢复的。这方面的改变包括：减少农药的使用；停止捕杀大型蛇类、蜥蜴、蛙类、龟类等；停止毁林行为，特别注意保护水沟边的植被；限制放牧等。要让老百姓不捕杀蛇类等野生动物可能是一件非常不容易做到的事情，但是通过适当的教育活动能够让群众认识到蛇等动物在控制害虫方面的价值。

鸟类

Douglas F. Stotz 权锐昌 李正波 Debra K. Moskovits

我们估计在高黎贡山至少有 600种鸟类，约占中国全部鸟类种数的一半。在9天的考察中，鸟类考察小组在保护区内一共记录到179种鸟类(附录4)：百花岭(东坡) 121种；大塘(西坡) 104种；赧亢(南端) 54种。据以前的文献(薛纪如等, 1995)记载，保护区一共记录到350多种鸟类，其中很多为低地鸟类(分布在海拔1,500m以下的地区)、水鸟或喜欢开阔地的种类，而这些种类在保护区内数量不多。在原名录的基础上，我们又发现了23种，其中18种为森林鸟类。新记录中有3个种的分布区有明显的拓展，包括 斑翅凤头鹃 (*Clamator jacobinus*) (在古丝绸之路上记录到的，原来认为只分布在西藏)，黑背燕尾 (*Enicurus immaculatus*) (在大塘和百花岭都观察到，原来只在腾冲记录到一次)和宽嘴鹟莺 (*Tickellia hodgsoni*) (在百花岭附近观察到，填补了西藏和滇东南之间的空白)。

保护区内高山(海拔1,500 m以上地区)鸟类异常丰富，估计约有350种森林鸟类(在保护区内已经记录到约225种)。根据考察中发现的新记录的数量，我们推测在保护区内还有100种左右的森林鸟类未被记录。估计来该地区过冬或从该地区迁徙的鸟类在30种左右(已经记录到16种)。在保护区周边地区的低地和受到频繁干扰的地区，目前至少已经记录到130种鸟类，估计今后的调查至少还能够记录到130种左右。

在高黎贡山海拔 2,400m以下的山坡上，鸟类物种特别集中，许多种类分布的地理范围非常狭小。即使在类似的海拔高度，三个考察地点的鸟类组成也有很大的差异。

在整个考察期间，只有在百花岭一线(沿着古丝绸之路从百花岭管理站一直到海拔3,100m左右的南斋公房垭口，图3))我们能够真正对沿海拔高度鸟类分布的情况进行评估。这条线路上鸟类的分布可以明显地划分出三个地带: 海拔1,500m-2,000m 2,000m-2800m, 2,800m-3,100m。没有发现哪个物种在所有海拔高度都有分布(进一步的研究可能会发现有个别物种的分布范围能够到达所有这些海拔高度)。在少量几个属，如雀鹛属(Alcippe)、凤鹛属(Yuhina)和太阳鸟属 (Aethopyga)，我们发现了垂直高度替代种。不过，这种替代并没有一般热带高山鸟类的替代现象明显。一般说来，高黎贡山的亚热带和温带鸟类的垂直分布范围要比美洲或非洲热带鸟类的分布范围广。某一海拔高度的鸟类的数量可以通过从其它海拔高度的鸟类的迁徙而得到补充。特别是，海拔2,000m左右的鸟类的丰富程度要靠海拔更低的一些地方迁来的鸟进行补充。如果保护区以外低海拔地区的森林(图2B, 2C, 3D)受到破坏或丧失，高海拔地区的物种多样性也会丧失。

高黎贡山记录到的森林鸟类中有四分之一分布范围狭小，所以很容易受到影响。高黎贡山有很多分布狭窄(分布范围不足50,000平方公里)的特有鸟类, 这些鸟类分别属于国际鸟类保护联合会确定的两个特有鸟区(EBA)—云南山地和东喜玛拉雅区 (Stattersfield et al. 1998)。在9天的考察中，我们观察到了5种这样的鸟类，包括2种分布狭窄的云南山地种类滇䴓 (*Sitta yunnanensis*) 和褐翅鸦雀 (*Paradoxornis brunneus*) (两种都是在大塘发现的)和3种东喜玛拉雅区种类，包括宽嘴蟆莺、楔头鹩鹛 (*Sphenocichla humei*) 和丽色奇鹛 (*Heterophasia pulchella*)。前三种为高黎贡山新记录。还有3种(白尾梢虹雉 (*Lophophorus sclateri*)、红腹咬鹃 (*Harpactes wardi*) 和 纹胸斑翅鹛 (*Actinodera waldeni*)) 东喜玛拉雅特有鸟区的种类在保护区也有记录，不过，我们此次考察中没有观察到。到目前为止，高黎贡山地区已经记录到8种分布范围非常狭小的鸟类。我们还记录到了37种(占此次考察观察到的鸟类总数的22%) 分布范围在100,000平方公里以下的鸟类。

国际鸟类保护联合会一直利用国际保护联盟(IUCN)的标准来对世界上的所有鸟类确定濒危等级。按照这个标准排序，我们发现在高黎贡山观察到的鸟类中有10种已经处于"受威胁"或"濒危"状况。先前在高黎贡山就已记录到的鸟类中有9种处于"受威胁"或"濒危"状况。这些鸟类的主要威胁因子是生境的丧失。至少有7种雉类(包括5种已经被国际鸟类保护联盟确定为"受威胁"种类)处于"易危"状况。目前，来自打猎的压力看来有所缓解，但是，应该对雉类的种群进行监测。我们必须立即采取行动，保证这些处于"接近受威胁"状况的鸟类不会进一步恶化为"受威胁"的状况，保证让目前处于"受威胁"状况的种类不至于变成"灭绝"状况。作为鸟类的避难所，高黎贡山具有十分重要的地位。从长远来讲，保护区的面积必须足够大，才能够真正维持必需的种群数量。

高黎贡山是各种生物地理成分汇聚的地方。除了前面所提到的该地区代表了两个特有鸟类分布区以外，它也是喜玛拉雅、华中山地、东南亚和古北区等各种鸟类区系成分的汇合之地。高黎贡山靠近画眉科 (Timaliidae) 鸟类的生物多样性中心。保护区内约有60种画眉亚科的鸟，我们在此次考察中一共观察到了43种。 画眉亚科鸟类具有各种各样的生态学特征、形状、个体大小和嘴部特征。在高黎贡山种数分布比较多的属有噪鹛属 (Garrulax，共记录到12种)、雀鹛属(共记录到7种)和凤鹛属(共记录到7种)。

关于季节性变化对高黎贡山的鸟类区系的影响，我们尚无法作出评估。考察时已经是大多数森林鸟类的孵化季节的末期，我们观察到了许多尚带着雏鸟的鸟类(附录4)。高黎贡山有许多鸟类(约35种)就在该地区过冬，也有许多鸟类在该地区繁殖但到别处去过冬(约45种)。此外，高黎贡山有许多鸟类随着季节的变化其活动范围在海拔高度上具有一定的变化。这种迁移方式意味着如果我们要成功地保护高黎贡山的鸟类，除了保护区本身以外，我们还要对其周边地区进行谨慎管理。

大型哺乳动物

艾怀森　蔺汝涛

在本次考察中，共调查到42种哺乳动物的直接或间接证据（见附录5），其中在百花岭地区调查到34种，大塘地区调查28到种，在郝亢调查到21种。估计在整个保护区生活有150余种各类哺乳动物。保护区动物区系复杂，主要为东洋界，也有一些古北界动物种类。保护区内特有种丰富，约4种; 森林兽类的比例也比较高，达85%以上。

考察中，我们主要根据直接证据如动物实体和叫声等和间接证据如粪便、足迹、觅食痕迹以及半结构式访谈等来进行调查。在对村民特别是猎人的访谈中，为了保证收集到的资料的可靠性，我们还专门向访谈对象展示了一些在保护区并不存在的动物的照片，以便寻找到可靠的访谈对象。

本次调查比较重要的发现就是在海拔2,000m附近地区有小熊猫(Ailurus fulgens)广泛分布，而过去认为该物种一般都分布在海拔3,000m左右的高山地带。考察中还发现该地区的白眉长臂猿 (Hylobates hoolock)、水鹿 (Cervus unicolor)等珍稀濒危动物都有一定数量的增长，

此外高黎贡山珍稀濒危动物丰富，被列入中华人民共和国重点保护的一、二类珍稀濒危动物保护名录的野生动物就有27种，在本次调查中发现了猕猴 (Macaca mulatta)、灰叶猴(Semnopithecus phayrei)、白眉长臂猿 (Hylobates hoolock)、短尾猴 (Macaca arctoides)、黑熊 (Selenarctos thibetanus)、小熊猫 (Ailurus fulgens)、金猫 (Catopuma temmincki)、水鹿 (Cervus unicolor)、羚牛 (Budorcas taxicolor)、鬣羚 (Capricornis sumatraensis)、斑羚 (Naemorhedus caudatus)、巨松鼠 (Ratufa bicolor)等13种珍稀濒危动物的踪迹或实体。

与其它调查对象一样，由于从海拔1,500m左右到山顶都是连片的森林，所以大型哺乳动物在海拔高度上的连续分布在百花岭表现得最好。虽然在大塘也有大片的森林，但是在许多地方特别是低海拔地区，人类的干扰一直存在而且比较严重。郝亢附近的森林遭受的人类活动的干扰也比较严重，林下植被已经很少。估计在这些地方个体较小的大型哺乳动物 (如灵长类、麂和野兔等) 可能比个体较大的哺乳动物(如穿山甲、黑熊和大型猫科动物等) 的数量会多一些。

威胁因子

高黎贡山国家级自然保护区保护着琳琅满目的自然群落。通过合理的规划，保护区是能够为自然保护和周边社区的协调发展创造具体条件的。但是，保护区还面临着许多来自内部和外部的压力。如果不采取措施缓解这些压力，就会影响保护区及其附近的人类和生物群落的稳定和长期存在。兹对该地区面临的一些主要威胁因子分析如下:

1. **低海拔地区生境的持续退化或完全丧失**

   保护区周边地区没有纳入保护范围，这些地方的生物资源正在受到极大的威胁。保护区的海拔低限为1,500 m左右，在周边地区的山坡和谷地，只剩下一些不连续的森林，而且大多数已经退化 (图2B，2C，3D)。但是，就是在这些退化、不断片段化的林子里，我们发现了许多珍稀、濒危或分布范围狭小的物种。正是这些低海拔山坡蕴藏着许多当地的特有物种，同时帮助维持着高海拔地区的动物种群。考察中发现的3个植物新种都是在大塘附近的低海拔林子里采集到的。

造成低海拔地区植被退化或完全破坏的主要原因有:

(1) 农业生产(烘烤烟叶是对森林植被最具破坏性的一种生产活动);
(2) 烧木炭;
(3) 毁林建牧场;
(4) 林下放牧(放牧活动在高海拔地区保护区内也存在);
(5) 薪材和建材消耗;
(6) 水土流失
(7) 农药及农用化学物质的使用(特别是对溪流、水塘等的破坏)

2. 小溪、河流受到干扰,敏感流域遭到破坏

许多动物种类包括一些高黎贡山特有的种类的生存直接依赖于小溪和河流。更多的物种对这些小溪和河流存在着间接的依存关系。在高黎贡山周围的农业区域,小溪、河流等已经几乎被破坏殆尽(河水被引作灌溉用水或被农药等污染)。源头地区的小溪流的破坏造成的后果尤其严重,因为高黎贡山两栖类动物的多样性主要就分布在这些源头地区。此外,湿地被污染也会影响大型真菌的多样性,进而影响植物的生长、植被的结构和动物的食源。

3. 牲畜进入保护区

牛、猪和羊等进入保护区会导致森林的迅速退化。除了对植被造成直接的伤害外,这些牲口还造成土壤压实,使真菌和其它生物无法生长。如果来自牲畜的压力长时间居高不下的话,首先会造成林下植被及其相关的一些动物的消失。

4. 污染物(如农药等)漂移进保护区(见 2.)

5. 动植物资源的过度利用

蘑菇的过度采集会对生态系统带来潜在的问题。必须对蘑菇的采集进行研究和管理,保证在每年都能够留下足够的子实体供下一年生长。同样,打猎也会危及到大型动物和鸟类特别是雉类的数量。虽然在保护区及其周边地区的集体林内不许打猎(集体林内

管理得稍松一些),但是,也有些地方没有得到很好的管理。应该对所有的打猎和采集活动都进行适当的管理和监测。

6. 来自创收的压力

在保护区内或周边地区开展创收活动如生态旅游活动的呼声为保护区与周边地区合作对资源进行有利于生态保护的利用提供了很好的机会。但是,如果不制定严格的监测计划对创收活动进行谨慎管理和适时调整,这些创收活动会因为资源的过度开发、缺少基础设施、污染和对自然环境造成干扰等对自然资源和人类发展都带来不良的后果。

7. 缺乏"绿色技术"方面的信息和获得这些信息的途径

农民生物多样性保护协会的成立意味着当地在社区参与自然保护方面已经迈出了一大步,但是百花岭的大多数村民缺乏有关环境安全的耕种方式和燃料来源的基本信息。当地村民对教育、医疗保健和卫生等非常关心,在与村民一同实现长期和短期的保护目标的时候必须对这些问题进行认真考虑。

保护对象

根据物种和群落的全球或区域稀有性、对群落动态的影响及其在生态过程中的重要性等标准，我们建议将下列物种和种群作为高黎贡山的重点保护对象：

生物群落保护对象	低山（海拔2,000m以下）森林（包括已经受到干扰的那些森林）； 罕见地质条件下的矮林及其它群落； 所有的原始森林及完整的垂直植被带谱； 东西坡比较有代表性的植被群落； 具有不同年龄结构和大量枯木的植被群落（以便维持真菌的多样性）。
植物保护对象	特有类群（高黎贡山特有种类）； 濒危物种，如 大树杜鹃（*Rhododendron protistum var. giganteum*）、 秃杉（*Taiwania flousiana*）、桫椤（*Alsophila spinulosa*）等； 药用植物。
真菌保护对象	食用或药用真菌； 与各植被群落和特定植物（如壳斗科和松科等）有关的菌根真菌； 起分解作用的真菌。
两栖及爬行动物 保护对象	以高山溪流（特别是3,000m以上）为栖息环境的两栖动物； 中山（海拔1,500m – 2,700m）亚热带种类； 高黎贡山特有或分布范围狭小的种类； 具重要经济价值的种类（如蜥蜴、蟒蛇、龟等）； 低海拔、受威胁的种类。
鸟类保护对象	山地常绿阔叶林鸟类（特别是在海拔2,000m以下的地区）； 以快速流动的溪流为生境的鸟类； 以针叶林为生境的鸟类； 大型林栖雉类（已知在高黎贡山有7种，其中3种处于"易危"状态、2种处于"接 近受威胁"的状态； 特有种类（特别是那些分布地理范围非常狭小的特有种类）； 处于"受威胁"或接近受威胁"的物种。

大型兽类保护对象	灵长类： 短尾猴（*Macaca arctoides*）、猕猴（*M. mulatta*）、 灰叶猴
	    （*Semnopithecus phayrei*）、白眉黑长臂猿（*Hylobates hoolock*）；
	大型食肉动物： 黑熊（*Selenarctos thibetanus*）、小熊猫（*Ailurus fulgens*）、
	    金钱豹（*Panthera pardus*）、孟加拉虎（*Panthera tigris*）；
	大型有蹄类动物： 水鹿（*Cervus unicolor*）、羚牛（*Budorcas taxicolor*）、
	    鬣羚（*Capricornis sumatraensis*）、斑羚（*Naemorhedus caudatus*）；
	珍稀、濒危种类。

高黎贡山国家级保护区为各种生物群落和大量分布范围狭小或濒危物种提供了庇护所。短短9天的考察就大大地丰富了保护区的物种名录，进一步说明：(1) 该地区在保护全球特有的生物资源的长期存在方面具有非常重要的意义；(2) 我们对保护区的认识还十分有限。

保护区为人类保护该地区特有的各种生态群落创造了条件，同时也为周边地区的多元文化与社区周围以及保护区内的生物多样性资源的管理和恢复协调发展提供了机会。通过使当地群众受益、激发当地人的自豪感、培养大自然好管家的精神和保存乡土文化，保护区能够成为国际上开展生态旅游的典范。兹将保护和保护区综合管理能够给该地区乃至全世界带来的一些主要好处介绍如下：

1) **高黎贡山国家级自然保护区是一个具有全球重要性的保护区 —— 从海拔1,500m左右的缓坡地带到 海拔4,000 m以上的峻峭山峰 —— 保护着东亚、喜玛拉雅和古北界的生态群落**
高黎贡山从低到高、从东到西分布着连片的森林，这些森林孕育着特有的动植物区系，保护着新物种形成的各种过程。

2) **使目前受到威胁的低海拔地区的丰富的动植物资源的生境得到恢复**
在目前的保护区边界以外的低海拔地区还分布着一些不连片的森林，但是这些森林所蕴涵的生物多样性非常丰富，生活着许多分布范围十分狭窄的动植物种类。对这些片段化的森林加以保护和拓展并辅之以适当的管理措施将能够使高黎贡山目前被人们忽略的大量的生物多样性的关键生境得到恢复。

3) **保护区有望成为成功的、以当地生态和文化为基础的、给保护区和当地社区都带来利益的生态旅游的典范**
在保护区周边地区分布着众多社区，这些社区是云南省独特的文化多样性的代表(图8)；游客和当地村民将有机会体验当地得天独厚的自然和文化多样性。通过精心设计、规划以及周边村落的全面参与，旅游活动能够为保护区的长远管理吸引资金，也能够让当地村民参与到有利于环境保护的经济活动中来。

4) 使周边社区与保护区人员在保护区的管理和利用方面开展成功合作

当地人能够成为保护区的守护者。只要在百花岭开展适当的培训活动,就能够实现提高周边地区村民的生活质量、减少自然环境的威胁和压力和为当地日益增长的旅游活动提供社会基础这三个互为补充的目标。

5) 使水域以及其它食用和药用资源得到保护

对保护区内资源的有效保护和保护区边界以下的地区的低地森林植被和动植物资源的恢复能够给周边社区和当地的经济带来直接的利益。山上的水域保护对于山下农业用水的持续供应举足轻重。部分村民采集自然资源用做食品、药物和增加收入。特别如菌类,一直是当地人的重要食物和现金收入来源。对大型真菌进行有管理的利用应该是高黎贡山可持续采集活动的一项重要内容。

6) 使保护区成为研究进化学 (特别是物种形成)、生态学 (特别是生物迁徙、生境利用和优势树种的生长) 和开展保护活动等的基地

保护区东坡连绵的森林和西坡保护区外的块状森林为研究保护和生物多样性问题如物种形成、生境利用和两栖动物减少的原因提供了极好的场所。

高黎贡山国家级自然保护区为高黎贡山南段（主要是海拔2,000m及以上地区）的动植物区系起到了至关重要的保护作用。不过，要发挥其作为全球特有的生物群落资源的潜力，必须拓展现有的保护区范围，将目前保护区边界以外但物种丰富的低海拔山地也纳入保护范围。我们认为应该 (1) 将从怒江一直到龙江的大片区域（图3）都纳入保护范围；(2)将保护区南北连接起来。此外，应该通过综合性的管理措施和生态兼容的经济活动来加强保护区周边地区的社区的发展。

要实现上述设想，需要采取下列关键性措施。为了使这些保护目标能够目的明确和重点突出，还需要开展积极的研究和调查工作。在附录7中我们还就建立生态旅游旅舍等提出了详细的建议。

保护和管理方面

1) **核心区严禁开展任何活动，仅开放部分区域用于科学研究**
   鉴于高黎贡山森林极其重要的生物学价值和周边地区人口对保护区施加的巨大压力，我们建议严禁在现在的核心区范围（主要是高海拔地区;见图3）开展任何形式的人为活动。

2) **加强对核心区以外的地区的保护和巡护管理**
   加强对低海拔地区（开展旅游活动的地区）的森林的保护对于维持森林的完整性具有至关重要的意义.

3) **将核心区尽可能延伸到低海拔地区**
   保护区的低海拔地区的动植物多样性极其丰富，包括许多分布范围狭小的易危物种。目前，由于保护区内正受到持续的人为干扰、保护区外的块状森林受到水土流失等的影响，这些生物多样性正濒临灭绝的危险（图2C）。

4) **将保护管理从怒江一直延伸到龙江边**
   目前，人们在讨论将保护管理一直从东面的怒江边延伸到西坡的龙江。不过，保护河谷地带的资源固然重要，但是，在确立保护级别的时候必须要充分考虑河谷地带30多万人口的生活问题。我们认为比较合适的方式是在现有保护区边界外建立"遗产区"，以允许当地群众继续利用当地资源，展现活生生的人文自然景观，同时又要建立激励机制，鼓励文化保护和生态恢复活动，鼓励有利于生态保护的可持续的经济发展活动。目前可以立即开展的一项活动就是保护区和社区合作对周边地区的自然资源进行管理，以便维持这些地方的文化和生物多样性。

5) **恢复和保护低海拔地区的森林植被；通过植被恢复等措施将目前分散的林块联结起来，并使之与更大的保护区联成一片**
   尽管时间短暂，但是，考察队却在低海拔地区发现了一些新种。所以为了保护低海拔地区那些具有全球重要性的生物多样性的长期存在，我们建议保护区管理部门与周边地区村民密切合作，共同恢复社区周围尚未被完全破坏的森林植被，利用乡土树种在宜林荒山开展造林活动以扩大动植物的有效生境。建议将保护区周边地区的所有现存森林包括赧亢附近的小块林地和大塘附近受到高度干扰的森林都纳入保护管理范围。

6) 加强百花岭村参与规划和实施生态旅游活动的能力

保证当地居民能够真正参与旅游活动的规划并从旅游开发中获益的一种方式就是成立一个由所有自然村的代表参加的、积极开展活动的村级生态旅游协会（类似于农民生物多样性保护协会）。该协会可以与村委会一道讨论和实施旅游计划和规章制度等。目前大鱼塘的旅游管理委员会的规模和覆盖范围都还太小，不能代表所有自然村的利益，所以还无法在更大的范围内起作用。

7) 在低海拔地区的山坡和谷地，研究和实施一些有利于生态保护的农业生产活动并开展生态恢复活动

鼓励农民发展多种经营，减少化肥和农药的使用，控制农用化学物质对环境的污染。此外，应该通过一定的机制使农民从生态恢复活动中得到好处。

8) 开发适合农民的支付能力的替代能源

目前，周边社区主要使用薪材作为燃料。开展旅游活动必然会增加对燃料的需求，从而增加保护区资源的压力。目前正在该地区推广使用一些替代能源如沼气等，但是当地村民仍然觉得成本太高。

生态旅游方面

1) 保证所有生态旅游活动能够给保护区和周边村民带来直接的利益

开展生态旅游活动以及其它生态兼容的经济活动的目的就是要为当地带来经济利益，而这种经济利益实现的前提是要有利于生物多样性和周边社区的文化和社会资源的保护。旅游开发要有利于保护和发扬当地的传统文化，要遵从社会文化自身的发展规律，还要充分尊重当地人的权利和隐私。

具体的一些建议包括 (1)使用当地导游；(2)利用当地食品；(3)门票收入中的相当一部分应该直接用于保护区的保护与维护，用于加强参与旅游活动的社区的能力；(4)旅游基础设施的修建要充分考虑生态保护。

2) 对保护区的游客承载量进行研究，并根据研究结果控制游客接待数量

入口处的旅舍和游客接待中心（附录8）是按照每天100人来设计的（估计保护区能够承载的最大容量也就是每天100人左右）。不过对此需要开展进一步的研究。

3) 旅游活动的开展和旅游基础设施的修建要尽量减少对敏感生物群落的影响，还要尊重和保存当地文化

无计划的开发将会危及到当地的自然资源和文化资源，而正是这些资源才使得该地区具有国际上独一无二的特点和风景秀丽的自然景观。要保护这些生物资源，必须对游客的数量和旅游活动进行严格管理（特别是古丝绸之路上）。要制定出科学的计划对旅游活动的环境影响进行监测并根据监测结果对旅游活动和基础设施的修建等进行相应的管理。当地村民应该参与旅游活动的设计和管理的整个决策过程，以使他们的隐私和生活质量能够得到充分的保护。

(1) **修建旅舍和游客接待中心**

修建这些设施时，要尽量减少对有限自然资源的消耗和对保护区的干扰，建筑风格要有当地特色，要充分依靠现有的建筑物如管理站等设施，要避免在保护区内修建任何新的设施（附录8）。

(2) **控制游客数量和旅游活动**

(3) **限制可能对生物群落造成损害的活动并进行监测**

— **对需要在保护区内过夜的旅游活动加以限制**

这类活动需要在保护区内修建设施、生火做饭、使用马帮等，对环境造成的影响最大。

— **将保护区内的旅游活动限制在维修得比较好的路上**

在保护历史遗迹（如丝绸古道）的前提下，应该修建旅游便道、对这些便道进行适当维修并要求游客使用这些便道进行旅游活动以便减少水土流失。

— **正确处理垃圾**

— **禁止或严格限制利用薪材做饭或取暖**

— **限制马帮的使用**

4) **将高黎贡山国家级自然保护区与滇西其它众多旅游景点联成一片**

这样既可以减 少旅游活动对保护区的压力又能丰富游客的旅游体验，同时还有利于提高社区 能力和保护本地文化和景观

(1) **制定区域性生态旅游规划，加强展示，使游客一到保山或芒市机场就能够享受到高质量的旅游体验。**

(2) **设计一系列到本地区其它旅游景点的一日游活动**

(3) **制定展示保护区的计划，对保护区的敏感区域进行分析，确定游客进入保护区的地点，从而尽量避免游客从一个地方进入保护区，减少环境影响。**

确定各个地区吸引游客、具有教育意义的景点和资源。

(4) **制定游客指南 和VCD等**

以鸟类和蛙类等为主题的VCD能够大大增加游客的体验。植物、真菌、两栖动物、爬行动物和周边地区的历史遗迹等则以照片或小册子等方式向游客介绍比较合适。

(5) **开展环境影响较低的野生动物观察活动，以增加游客体验**

例如，在低敏感区域，开辟小块（2-3亩）、浅的水塘，修建简单的设施（游客可以小坐和避雨的地方），游客可以在这些地方聆听蛙鸣、观察蛙和蝾螈等交配或观看前来饮水的动物等（赧亢附近植物园边的水塘就是个很好的地方）。

(6) **改善当地群众的生活条件、提高游客的接待水平，加强百花岭村的社会、文化资源和历史遗迹.**

5) 提高社区群众的能力, 使百花岭村的村民能够有效地参与旅游开发并从中受益

除了加强和提高社区能力的措施以外, 还应该对当地群众进行培训并提供适当的帮助使他们能够参与到旅游开发中去。例如, 应该对导游和与游客打交道的村民进行培训, 培训内容包括外语 (英语、日语等)、解说技巧、历史知识、生物多样性和民族多样性等。此外, 还必须对旅游接待系统如餐饮业、运输业、商业网点等的服务人员进行培训。所有雇员, 无论其具体职责, 都应该接受有关当地的自然景观和文化资源的培训。导游和其它旅游从业人员必须从当地社区挑选、熟悉当地的文化和传统、而且必须保证能够长期在当地工作。应该让当地村民成为旅游活动的真正利益相关者, 而不仅仅是简单的打工挣钱的人。要转变村民与游客之间的关系, 使之形成主人和客人的关系, 而不是简单的服务者和顾客的关系。如果能做到这样, 就能够使村民对游客的热情出于自然、发自内心, 使游客得到真正的体验。

研究方面

1) 弄清农药的使用对两栖动物种群的影响

例如, 可以在大塘建立两个面积在10-20公顷左右的稻田作为实验点。实验田内5-10年内不使用农用化学物质。该项目可以取得一举多得的效果:

(1) 为人们弄清使用杀虫剂等农业化学物质对生活在这些农田中的两栖动物的影响提供了极好的机会;

(2) 让当地村民参与到保护自然遗产的过程

(3) 人们可以比较杀虫剂和蛙类在灭虫中的作用 (农田中的蛙类能够消灭掉大量的、能够造成巨大经济损失的害虫);

(4) 利用该试验, 人们可以开展农药杀虫和蛙类灭虫的成本效益分析 (要开展该项科研活动, 需要给农民提供适当的资金扶持)。

2) 确定采集野生菌类对生物多样性的影响, 制定相应策略, 使当地群众能够可持续地利用这些菌类资源获取收入

一旦弄清了采集活动对生物多样性的影响关系, 就可以制定相应的管理和监测计划对采集活动进行管制。

3) 确定森林鸟类和高山灌丛鸟类的海拔分布和丰富度随季节变化的情况

虽然许多鸟类都有在不同海拔高度游移的现象, 但是, 人们对此的认识却非常有限。随着低海拔地区的生境的丧失, 这方面的认识对于鸟类区系的成功保护就显得十分重要。在同纬度的其它热带地区, 鸟类在不同海拔高度上游移的现象在食果鸟类和食蜜鸟类中比较明显, 在种子传播和植物授粉方面具有一定的意义。

4) 确定中山森林鸟类的生境利用模式

这方面的研究对于管理具有一定的意义

5) **确定受威胁的大型哺乳动物的基本生态学特点并制定相应的保护策略**
研究的重点对象包括：短尾猴(*Macaca arctoides*)、猕猴(*M. mulatta*)、灰叶猴 (*Semnopithecus phayrei*)、白眉长臂猿(*Hylobates hoolock*)、黑熊 (*Selenarctos thibetanus*)、小熊猫(*Ailurus fulgens*)、 金钱豹 (*Panthera pardus*)、孟加拉虎(*Panthera tigris*)、水鹿(*Cervus unicolor*)、 羚牛(*Budorcas taxicolor*)、鬣羚(*Capricornis sumatraensis*)和斑羚 (*Naemorhedus caudatus*)。

6) **对植物，两栖、爬行动物开展分子生物学层次上的系统研究，以便认识地理变异水平和可能的隐秘种类。**

7) **对该地区的蕨类和大型真菌进行详细的分类学研究**
高黎贡山已经报道的真菌中有许多种类与欧洲和北美的许多真菌同名，但实为不同的物种，甚至是一些新的物种。这一点对于保护具有重要的意义，因为这些物种中有许多实际上是高黎贡山地区（至少是亚洲）的特有种，因此，维持这些种类的种群水平就显得十分重要。此外，由于高黎贡山兼有东、西、南、北的种类，所以，高黎贡山的真菌资源对于认识大型真菌的生物地理学具有十分重要的意义。

8) **研究重要树种的物候学特征和果实生长; 研究经济林木的自然增长和生长情况; 研究侵入物种的传播情况**
这些研究对于制定适当的管理计划具有一定的意义。

9) **进一步研究当地村民的文化和社会传统以便更好地认识其文化、社会与自然遗产**
对于百花岭村及保护区周边地区的村子进行进一步深入研究，收集基础资料，以便对当地居民和游客进行有的放矢的教育，为文化保护活动提供依据。此外，对明代有关高黎贡山地区的文献和记载进行考证和对古丝绸之路上的一些历史遗迹进行考古学研究将会大大丰富人们对该地区的历史和重要事件的认识。

进一步调查方面

1) **深入调查该地区的文化、历史资源**

2) **对受到干扰的、低海拔地区的重要生物进行调查，确定恢复和保护工作的重点地区**

3) **通过抽样调查，估计关键物种的种群大小**
对于那些容易受到干扰的物种来说，这种调查尤为重要。了解种群大小有利于制定科学的管理决策。

4) 开展生物地理和分类学方面的补充调查

调查主要集中在以下方面:

(1) 爬行动物特别是高海拔地区的蛙类和保护区内所有的蛇类调查

实际上,对蜥蜴和蛇类进行抽样调查的唯一有效的方法是使用带有网兜的铝制漂流网.

(2) 调查海拔3,000m以上的鸟类资源

(3) 进一步的植物调查

人们对距离百花岭管理站和古丝绸之路较远的地区的植物的认识还十分有限。特别地,对于土壤条件和降水情况与东坡具有明显差异的西坡应该进行更加详细和深入的调查。

(4) 真菌调查

高黎贡山的真菌多样性目前还只有一小部分得到记录。采用随机和定量的调查方法能够进一步揭示真菌群落的物种组成、分布格局以及潜在的宿主、地点和分布区的土壤条件等。

监测方面

1) 评估生态旅游活动的效果

需要评估的内容包括旅游活动对生物和文化群落的影响、旅游活动对保护区和周边地区群众带来的好处、旅游活动的可持续性等。要成功地开展监测活动,必须让当地群众参与到监测计划的制定和实施中来。

2) 通过定期调查监测雉类种群大小

偷猎目前似乎还没有对保护区内雉类种群构成威胁,但是,从长远来讲,可能会成为一个威胁。

3) 对受到人为干扰的缓冲区的林鸟进行监测

4) 对两栖动物和珍稀植物进行监测,确定其种群变化趋势

如果监测结果证明保护区内两栖动物和珍稀植物的种群大小在下降,就需要进一步弄清楚下降的原因并制定管理措施。

5) 对一些具有食用或经济价值的大型真菌的种群(多度、子实体的大小和形成子实体的季节)进行调查

利用调查结果对野生真菌的采集活动进行管理,以便保护这些具有重要经济价值的真菌种群。

6) 监测"受威胁"的兽类种群的变化

ENGLISH CONTENTS

(for Color Plates, see pages 17-28)

FIELD TEAM

Gerald W. Adelmann (coordinator)
Openlands Project, Chicago, Illinois, U.S.A.
jadelmann@openlands.org

Huaisen Ai (mammals)
Gaoligongshan Nature Reserve Management Bureau
Baoshan, Yunnan, China
GLGS@BS.YN.cninfo.net

Lilan Deng (botany)
Southwest Forestry College, Kunming, China
gsfan@public.km.yn.cn

Victoria Drake (social and cultural assets)
Openlands Project and The Field Museum
Chicago, Illinois, U.S.A.
jecvcd@earthlink.net

Robin B. Foster (ecology)
Environmental and Conservation Programs
The Field Museum, Chicago, Illinois, U.S.A.
rfoster@fieldmuseum.org

Shangyi Ge (fungi)
Gaoligongshan Nature Reserve Management Bureau
Baoshan, Yunnan, China
GLGS@BS.YN.cninfo.net

Ken Hao (coordinator)
Columbia University
Center for U.S.-China Arts Exchange
New York City, U.S.A.
kmh101@columbia.edu

Peter J. Kindel (gateway lodge study)
Skidmore, Owings & Merrill, Chicago, Illinois, U.S.A.
peter.j.kindel@som.com

Zhengbo Li (birds and coordinator)
Gaoligongshan Nature Reserve Management Bureau
Baoshan, Yunnan, China
GLGS@BS.YN.cninfo.net

Rutao Lin (mammals)
Gaoligongshan Nature Reserve Management Bureau
Baoshan, Yunnan, China
GLGS@BS.YN.cninfo.net

Shiliang Meng (ecology)
Gaoligongshan Nature Reserve Management Bureau
Baoshan, Yunnan, China
GLGS@BS.YN.cninfo.net

Debra K. Moskovits (coordinator and birds)
Environmental and Conservation Programs
The Field Museum, Chicago, Illinois, U.S.A.
dmoskovits@fieldmuseum.org

Gregory M. Mueller (fungi)
Department of Botany
The Field Museum, Chicago, Illinois, U.S.A.
gmueller@fieldmuseum.org

Jiali Qin (ecology)
Southwest Forestry College, Kunming, China
qinjiali@lol365.com

Ruichang Quan (birds)
Kunming Institute of Zoology, Kunming, China
quanre@163.com

H. Bradley Shaffer (amphibians and reptiles)
University of California, Davis, C.A., U.S.A.
hbshaffer@ucdavis.edu

Xiaochun Shi (botany)
Gaoligongshan Nature Reserve Management Bureau
Baoshan, Yunnan, China
GLGS@BS.YN.cninfo.net

Douglas F. Stotz (birds)
Environmental and Conservation Programs
The Field Museum, Chicago, Illinois, U.S.A.
dstotz@fieldmuseum.org

Adam Thies (gateway lodge study)
Skidmore, Owings & Merrill, Chicago, Illinois, U.S.A.
adam.thies@som.com

Anne Underhill (social and cultural assets)
Department of Anthropology
The Field Museum, Chicago, Illinois, U.S.A.
auhill@fieldmuseum.org

Tianchan Wang (amphibians and reptiles)
Gaoligongshan Nature Reserve Management Bureau
Baoshan, Yunnan, China
GLGS@BS.YN.cninfo.net

Jun Wen (botany)
Department of Botany
The Field Museum, Chicago, Illinois, U.S.A.
jwen@fieldmuseum.org

Bin Yang (fungi)
Southwest Forestry College, Kunming, China
yangbinyb@hotmail.com

Jianmei Yang
Tourism and Planning Research Center
Yunnan Normal University
Kunming, Yunnan, China

Shaoliang Yi (coordinator)
Southwest Forestry College, Kunming, China
a1234567@public.km.yn.cn

Yu Zhang (amphibians and reptiles)
Southwest Forestry College, Kunming, China
swfczhangy@163.com

Xiaodong Zhao
Gaoligongshan Nature Reserve Management Bureau
Baoshan, Yunnan, China
GLGS@BS.YN.cninfo.net

Yuming Zhu (social and cultural assets)
Gaoligongshan Nature Reserve Management Bureau
Baoshan, Yunnan, China
GLGS@BS.YN.cninfo.net

COLLABORATORS

Baihualing Village Association
Baoshan Prefecture, Yunnan, China

Municipal Government of Baoshan City
Baoshan, Yunnan, China

The Field Museum

The Field Museum is a collections-based research and educational institution devoted to natural and cultural diversity. Combining the fields of Anthropology, Botany, Geology, Zoology, and Conservation Biology, museum scientists research issues in evolution, environmental biology, and cultural anthropology. Environmental and Conservation Programs (ECP) is the branch of the museum dedicated to translating science into action that creates and supports lasting conservation. With losses of natural diversity accelerating worldwide, ECP's mission is to direct the museum's resources—scientific expertise, worldwide collections, innovative education programs—to the immediate needs of conservation at local, national, and international levels.

The Field Museum
1400 South Lake Shore Drive
Chicago, Illinois 60605-2496 U.S.A.
312.922.9410 tel
www.fieldmuseum.org

Center for United States-China Arts Exchange

The Center for United States-China Arts Exchange at Columbia University was established through a formal agreement with China before the normalization of relations between the two countries and has been carrying out major projects since 1979. Founded by Chou Wen-chung, Fritz Reiner Professor of Musical Composition (Emeritus), the Center has sponsored programs, partnerships and collaborations in the arts, education and conservation that have included participants from the entire Asian-Pacific region as well as the United States and Europe.

The main focus of the center since 1990 has been a broadly defined cultural and environmental conservancy program for the indigenous peoples in Yunnan Province, known as the minority nationalities. Designed to create a comprehensive strategy for the continuation of their traditional cultures and for the preservation of their unique ecology, this program has involved hundreds of specialists from China, the United States, Europe and Asia, and has mobilized thousands of local cultural and environmental workers.

Center for United States-China Arts Exchange
423 West 118th Street, #1E
New York, New York, 10027 U.S.A.
us_china_arts@yahoo.com

Southwest Forestry College

Southwest Forestry College (SWFC) is located in Kunming, China and is under the joint administration of the Provincial Government of Yunnan and the State Forestry Administration. SWFC is the only institute of higher learning in forestry located in the western provinces of China. Its major tasks are to foster forestry professionals, to carry out forestry research and to provide technical support, especially in the southwest region. Currently, over 6,000 students, including a high percentage of ethnic minorities, are studying in the college's 13 departments and 10 research institutes. Major fields of study of the college include forestry science, resource management, rural planning, wildlife and nature reserve management, water and soil conservation, ecotourism, and wood processing. SWFC is actively involved in completing inventories to promote or establish nature reserves in Yunnan Province.

Southwest Forestry College
Bailongsi, Kunming, 650224
Yunnan Province, P.R.China
www.swfc.edu.cn
oicswfc@public.km.yn.cn

Gaoligongshan National Nature Reserve Baoshan Management Bureau

The Gaoligongshan National Nature Reserve Baoshan Management Bureau was established in 1993. It has jurisdiction over approximately 25% (99,675 hectares) of the reserve. It is responsible for: (1) protection and maintenance of the reserve, including law enforcement; (2) education and research, in partnership with outside institutions; and (3) sustainable development of ecotourism. The bureau has management centers in Longyang District and Tengchong County, as well as eleven management stations and two forestry police precincts. Its staff includes forestry professionals, technical specialists, and rangers who patrol the reserve.

Gaoligongshan National Nature Reserve Baoshan Management Bureau
No. 2. Park Street, Baoshan City
Yunnan Province, P.R. China
GLGSET@BS.YN.cninfo.net

Openlands Project

Openlands Project is a private non-profit organization that protects, expands and enhances open space—land and water—to provide a healthy natural environment and a more livable place for all the people of the Chicago metropolitan region. Founded in 1963, Openlands has helped ensure the preservation of over 50,000 acres of parks, natural habitat, forest preserves, bicycle trails, wetlands, urban gardens and places to observe nature. In 1982, Openlands founded the Canal Corridor Association, which led the initiative to create America's first National Heritage Corridor.

Gerald Adelmann, Executive Director of Openlands Project, also serves as a Board member of the Center for United States-China Arts Exchange, and brought together the Center with the Field Museum and volunteers from Skidmore, Owings and Merrill to visit the Gaoligongshan region.

Openlands Project
25 East Washington, Suite 1650
Chicago, Illinois 60602 U.S.A.
312.427.4256 tel
info@openlands.org

The Yunnan Provincial Association for Cultural Exchanges with Foreign Countries

The Yunnan Provincial Association for Cultural Exchanges with Foreign Countries is the largest non-governmental organization in Yunnan engaged in a wide spectrum of exchanges with many countries. Its goals are to make Yunnan better known internationally, to enhance friendships with the peoples and the nations of the world, and to exchange knowledge and experience through visitation and collaborative projects.

Yunnan Provincial Association for
Cultural Exchanges with Foreign Countries
Milesi Street, Kunming, 650021
Yunnan Province, P.R. China

Skidmore, Owings & Merrill LLP (SOM)

Founded in 1936, SOM is one of the world's leading architecture, urban design, engineering, and interior architecture firms. Since its founding, SOM has completed more than 10,000 projects in more than 50 countries around the world. The firm has received over 800 awards, including the first Firm Award from the American Institute of Architects in 1961.

As a concerned leader within the environmental design community, SOM has sustained a commitment to creating great and lasting projects for both commercial and non-profit clients. Through partnerships with both governmental agencies and non-profit advocacy groups, SOM has brought both conceptual and technical expertise to a variety of national and international pro-bono work including the Yunnan Initiative.

Skidmore, Owings & Merrill, LLP
224 South Michigan Avenue, #1000
Chicago, Illinois 60604 U.S.A.
www.som.com

This report is a demonstration project of the Yunnan Initiative of the Center for United States-China Arts Exchange at Columbia University. In conjunction with its Chinese Committee of Specialists, the Yunnan Initiative brings international experts to Yunnan to collaborate with local professionals, leaders, and citizens on recommendations for sustaining and protecting the province's remarkable environments and cultures.

THE GUIDING PRINCIPLES OF THE YUNNAN INITIATIVE ARE

Conservation Development should proceed without damage to culture, ecology, and society.

Inclusion Development and conservation strategies must be inclusive of all nationalities, and must build on local cultural heritage, in keeping with Yunnan's policy to develop the province into "a great province of nationalities cultures."

Education Public awareness of cultural and environmental values must become commonplace to ensure the long-term success of sustainable programs.

Tourism Development of tourism, which is a potential engine for economic development, must enhance Yunnan's culture and ecology, and must provide direct social and economic benefits to indigenous people.

Collaboration Strategies to integrate conservation and development should build on local, regional, national and international collaboration.

THESE PRINCIPLES GUIDE THE RECOMMENDATIONS IN THIS REPORT, WHICH SUMMARIZES THE FIELDWORK OF AN INTERDISCIPLINARY TEAM OF SCIENTISTS, ARCHITECTS, PLANNERS AND HERITAGE TOURISM EXPERTS IN JUNE AND JULY 2002. IT JOINS A REPORT ON THE WEISHAN HERITAGE VALLEY (CENTER FOR UNITED STATES-CHINA ARTS EXCHANGE, 2001) AS A YUNNAN INITIATIVE DEMONSTRATION PROJECT.

ACKNOWLEDGEMENTS

Since 1990, Professor Chou Wen-chung, Founder and Director, Center for United States-China Arts Exchange and Fritz Reiner Professor Emeritus of Musical Composition, Columbia University, has provided unwavering vision for conservation in Yunnan province. His leadership built the foundation for these rapid biological and social inventories in Yunnan, and continues to set the stage for conserving and reinvigorating Yunnan's indigenous resources.

Working with Professor Chou to lead the Yunnan Initiative are Gu Poping, Director of the Yunnan Association of Social Sciences, who has opened the door for international collaboration in Yunnan, and Professor Fan Jianhua, General Office Yunnan Provincial Committee. Professor Fan's incomparable knowledge of the people, cultures and history of Yunnan, his boundless energy, and the respect he commands, make him the linchpin of the Yunnan Initiative.

The Gaoligongshan rapid inventory teams received invaluable support from Mr. Yang Wenhu, Director, Department of Information and Publicity, Baoshan Municipality, Yunnan Province. He provided indispensable municipal support, and organized and participated in the Gaoligongshan Rapid Inventory Findings Briefing Conference.

Professor Yang Yuming, Vice President and Professor, Southwest Forestry College provided unparalleled knowledge of the botany of Gaoligongshan. His recognition of the importance of international cultural and scientific collaboration made him the ideal person to lead the Chinese participants, including many of his junior academic colleagues and graduate students.

Professor Wu Deyou, Southwest Forestry College, took up the unenviable but essential task of overseeing the organization and preparation of the Southwest Forest College rapid inventory teams. His careful and meticulous planning made the collaborative inventory proceed as planned under trying field conditions.

Professor Guo Huijun, Deputy Director, Xishuangbanna Tropical Botanical Gardens, is one of the top international experts in the conservation and botany of Yunnan. His briefing to the American teams on Yunnan and Gaoligongshan was invaluable.

Mr. Li Zhengbo, Deputy Director, Yunnan Gaoligongshan National Nature Reserve, provided tremendous assistance with logistics. We thank Mr. Li and all local park managers for sharing their deep knowledge of the region with all of us. All Chinese scientists were extraordinarily gracious and forgiving of the lack of Chinese language skills in their U.S. counterparts. Dr. Jun Wen was extremely generous with translations and constant problem solving.

Bringing the Field Museum into the Yunnan initiative would not have happened without the leadership of John W. McCarter Jr., who made an initial exploratory visit and continues to provide enthusiastic support.

Alaka Wali, Director of The Field Museum's Center for Cultural Understanding and Change, made it possible for the social asset inventory to take place, both financially and thanks to her pioneering research. She contributed her expertise from conducting research of this kind in South America and Illinois by advising the Baihualing Village study team.

Robin Groesbeck of The Field Museum provided helpful advice on possible exhibition ideas and curator Ben Bronson advised on possible exhibitions and research. International heritage tourism expert Cheryl Hargrove, of the Heritage, Tourism and Communications (HTC) Group, provided important guidance and advice to the team.

Thanks to the support and leadership of Philip Enquist, AIA, Skidmore Owings & Merrill provided invaluable pro bono services to recommend a design concept for an ecologically sensitive lodge and visitor center.

Funding for these rapid inventories came from The John D. and Catherine T. MacArthur Foundation, the Center for United States-China Arts Exchange, and The Field Museum.

Dates of field work	Biological inventory: 17 - 26 June 2002; social inventory: 1-14 July 2002
Region surveyed	Three areas in Gaoligong Mountain National Nature Reserve, at the southern end of the Gaoligongshan range (Baoshan District, Yunnan Province, China) at the border with Myanmar: (1) the Baihualing Station on the eastern slope of the Gaoligong mountains, along the Southern Silk Road, from 1,500 m to 3,100 m at the pass; (2) Datang, on the western slope, between 1,850 m and 2,700 m; and (3) Nankang, a pass at the south end of the reserve, between 2,000 m and 2,200 m (figure 3).
Human communities surveyed	Eight hamlets in the Baihualing village: Hanlong, Dayutang (Upper, Lower), Bangwai-Guxingzhai, Taoyuan, Laomengzhai-Baihualing-Malishan, Manggang, Manghuang. These hamlets directly abut the perimeter of the Gaoligong Mountain National Nature Reserve and form the primary gateway for access to the reserve from the east (figure 3).
Organisms surveyed	Vascular plants, macrofungi, amphibians and reptiles, birds, large mammals
Highlights of results	The Gaoligongshan region is ecologically unique; a major crossroads of north and south, east and west, temperate and subtropical. The continuous belt of forest from east to west and over the crest of the mountains provides an unparalleled opportunity to conserve the spectacular mix of ecological communities and also to maintain conditions that both create new species (mountain ranges separated by deep valleys) and prevent the extinction of old ones (e.g., absence of drought). In nine days in the field, the biological team found an extremely rich fauna and flora, with several new records for the reserve. Outside the reserve, the forest is almost gone.
	The human communities living at the foothills reflect the enormous cultural diversity in Yunnan. In 14 days in the field, the anthropological team identified resources and capacities in the hamlets immediately adjacent to the reserve on the eastern slope. These assets will serve as entry points for working with the communities to develop economic activities, such as ecotourism, that are compatible with the local ecology and culture. We summarize the highlights of our results below.
	Social resources and capacities: Our brief survey, the first of its kind in the area, focused on Baihualing village, one of the 109 villages documented in Gaoligongshan (Baoshan Management Bureau, 2002). We estimate that 2,100 people and more than 450 families live in Baihualing. Six ethnic groups are represented: Han, Lisu, Bai, Dai, Yi, and Hui. As a whole, Yunnan province has 25 out of the 55 ethnic minority groups recognized in China (Population Census Office of Yunnan, 1992). Expression of ethnic identity through native dress,

crafts, language, festivals and customs is not as pronounced here as in other areas of Yunnan. The principal income (up to 85%) of this agrarian society comes from cultivating sugarcane, rice, and coffee. Fuel is an overriding concern: we were told that over 90% of families have adopted the new fuel-efficient stove, resulting in over 60% savings in energy. We found that each hamlet has distinct social assets. Cultural assets include historic sites dating from the Ming dynasty up to World War II. Modern assets include the ability to produce traditional crafts such as embroidered shoes (figure 8E), wicker stools and traditional foods. Other assets include the presence of local organizations, such as the Gaoligongshan Farmers' Biodiversity and Conservation Association and its demonstration and pioneer families, a very small tourism organization in Dayutang, and a women's association in Manghuang. Education is a pervasive priority of residents.

Vegetation and flora: The botany team identified ca. 1000 species of plants during the nine-day field survey, collected about 300 species, and photographed 250 (figure 4). About ten percent of the flora is endemic, i.e., it occurs only in Gaoligongshan. The variation in flora among sites is substantial. Datang, on the geologically distinct west slope, has a remarkably different flora. We found at least three species new to science (two Araliaceae, one Vitaceae). The diversity of ferns (pteridophytes) is also high (Appendix 2); we recorded a number of new species and genera for Gaoligongshan.

Macrofungi: The mycology team observed over 200 species of macrofungi (between 1,500 m – 2,400 m) and collected vouchers of about 150 (Appendix 1). Only 22 of the 200 species had been recorded previously in Gaoligongshan. We found a number of north temperate species mixed in with species from tropical Asia and species endemic to China. And we found species with disjunct distributions in eastern North America and eastern Asia. Macrofungi, crucial for the maintenance of high-quality natural communities, are also an important component of the local diet and the local market.

Amphibians and reptiles: The herpetology team found seven species of snakes, four species of lizards, one species of salamander and 15 to 21 species of frogs (pending confirmation of identifications; see figure 5; Appendix 3). Among these findings, three species of snakes and two of frogs were new records for the region, and one high-elevation frog is endemic to Gaoligongshan. We found the abundance of common amphibians to be 2-10 times higher in the reserve than in the rice patties near Datang (figure 3D). The use of chemicals in agricultural areas around the reserve has a severe detrimental impact on frogs.

Birds: The ornithology team found 179 species of birds (Appendix 4) during the nine-day survey, of which 23 were new records for the region. Gaoligongshan has a rich forest avifauna, especially below 2,400 m. The elevation turnover of individual species is extensive, but we found no sharp differences in communities across elevation. We registered 43 species (25% of our total) with restricted ranges, including species representing two distinct endemic bird areas (EBA), the Yunnan Mountain and Eastern Himalayan EBAs. The reserve's current bird list includes at least 19 species that are threatened or near-threatened with extinction.

Large mammals: The mammalogy team registered 42 species (Appendix 5) through direct sightings and indirect evidence (tracks, scat, local interviews). Of these, 13 are nationally protected species, including four primates and the lesser panda. One of our significant findings was evidence of lesser pandas down to 2,000 m, well below the 3,000 m level to which they are typically thought to be restricted.

Main Threats

The primary direct threats to Gaoligong Mountain National Nature Reserve are (i) agricultural activities—including the use of chemical fertilizers—along the lower edge of the reserve (with associated disruption of streams and rivers and drift of pollutants); (ii) continued expansion of crop, pasturelands, and grazing into the reserve; and (iii) local needs for fuel given the few affordable alternatives to burning wood. Lack of basic information on environmentally safe alternatives to current farming practices threatens to extend these damaging activities into the future. Deforestation of the lower slopes places an enormous diversity of plants and animals—many of them restricted to the region—at risk of extinction. Eventual disappearance of these lower-elevation species would affect the dynamics of higher-elevation communities protected inside the reserve. Finally, the introduction of ecotourism in the region, while a tremendous opportunity, will threaten the reserve's integrity if not developed and managed carefully, with strict attention to the vulnerability of both natural and human communities.

Current Status

Gaoligong Mountain National Nature Reserve protects 405,549 hectares of the higher (upper and mid) slopes in the southern range of Gaoligongshan. The lower edge of the reserve varies from 1,500 to 2,500 m. The highest areas have been designated as an inviolate core, with no visitors allowed (figure 3). The exception is along the Southern Silk Road, which has been placed outside of the core area and allows visitor access to the highest elevations in the reserve. Land below the reserve boundary receives no formal protection and is a mixture of small-scale croplands, pastures, and disturbed forests (figures 2B, 2E, 3C, 3E). In 1994, the Chinese Ministry of Forestry allotted 8,550 hectares in the Gaoligong Mountain National Nature Reserve (6.8% of the total; all outside the core area) for tourism development.

Principal recommendations for protection and management

1) *Extend conservation management beyond the Gaoligong Mountain National Nature Reserve, from river to river (figure 3).* Expand the limits of the reserve, as possible, down the mountains to reach the lower slopes (figures 2B, 2C, 3D). Here an enormous array of plants and animals will face extinction with the continued conversion of remaining pockets of unprotected forests to agriculture. Beyond the reserve, collaborative programs with neighboring villages for ecologically compatible economic activities would stretch the effective area of conservation from the Nujiang to the Longchuanjiang Rivers, protecting both lower-slopes and highland communities.

2) *Keep the core of the reserve untouchable, with a few areas open to researchers; extend the core area to lower elevations where possible.* Because of the strong human pressure all around the reserve, we recommend that a significant portion of the reserve (the "core" area, see figure 3) remain completely off-limits, as it is now, and that the core be extended to cover lower elevations wherever possible.

3) *Restore and protect remaining lower-slope forests at the base of the reserve; extend currently isolated forest patches eventually to link one to the other and to form conservation corridors among the larger protected areas.* For the long-term survival of global biological treasures in the lower-slope forests, we recommend developing collaborative programs with the neighboring villages (figure 3B) to restore degraded patches of forest and to reforest (with native species) denuded stretches surrounding forest islands to increase and connect available habitat.

4) *Strengthen Baihualing village's infrastructure and capacity.* One opportunity to ensure that local residents are involved in planning and benefit from tourism activities would be to establish a vigorous village ecotourism association (in the fashion of the Farmers' Biodiversity and Conservation Association) representing the eight hamlets. This association would work with the existing village committee to discuss and implement plans and policies related to tourism. To be successful, this tourism association would function at a larger scale than the existing small committee in Dayutang, and would represent all hamlets.

5) *Research and implement less ecologically damaging agricultural practices in the valleys and seek opportunities for ecological restoration.* Increase options for farmers to diversify crops and to reduce use of polluting fertilizers and pesticides (which are also extremely expensive).

6) *Increase affordable options for fuel.* Currently, wood is the primary source of fuel. Tourism will increase pressure on the forests by increasing demand for fuel. Local villagers are unable to afford alternatives such as methane.

Principal recommendations for ecotourism	1) *Ensure that revenues from ecotourism activities directly benefit the reserve and the neighboring villages.*
	2) *Research carrying capacity for visitors in the reserve and carefully manage visitor loads accordingly (see Appendix 8).*
	3) *Design ecotourism activities and infrastructure to minimize impact on the sensitive biological communities and to strengthen neighboring villages; keep infrastructure outside the reserve.* Create a Gateway Lodge and Visitor Center—which builds on existing infrastructure and is compatible with local design traditions—as the headquarters for strictly managed tourism (see Appendix 8).
	4) *Limit and monitor activities that can damage biological communities.* Principal measures include (i) proper disposal of waste, (ii) no use of firewood for cooking and heating, (iii) restriction of activities to well designed trails (which are managed for minimal erosion), (iv) minimal use of pack animals, (v) limited overnight trips.
	5) *Approach the Gaoligong Mountain National Nature Reserve as one of a constellation of tourism destinations within this part of Yunnan Province.* This will reduce pressure on the reserve and will create a rich visitor experience, while strengthening communities and preserving indigenous cultures and landscapes.
Long-term conservation benefits	1) A globally important nature reserve—from the lower slopes up to the rugged crests at 4,000 m—protecting a unique mixture of biological communities.
	2) Restored habitat for a remarkable diversity of lowland plants and animals—many of them restricted to the region—that are currently at risk; sustained management of locally valuable natural resources.
	3) A replicable model for successful ecotourism that is ecologically and culturally sensitive, brings direct income to the local villages and to the nature reserve, is a collaborative project managed by the local villages, and introduces Chinese and foreign visitors to the biological and cultural riches of the region.
	4) Integrated management between the reserve and the surrounding villages to implement practices that protect the watersheds and reduce the use of damaging chemicals.

Why Gaoligongshan?

A lush crossroads of north and south, east and west, temperate and subtropical, the Gaoligong Region in southwestern China, at the border with Myanmar (Burma), is a unique blend of biological realms and biogeographic provinces (figure 2). The continuous belt of forest from east to west over the crest of the Gaoligong mountains provides pathways for an extraordinary mix of the flora and fauna from the Himalayas, the Palearctic, and the tropical elements of the Oriental realm of southeastern Asia. The conditions in Gaoligongshan that allow for a complete transition from temperate to tropical forests are unparalleled in the world. The region is a primary center of diversity and endemism and a top priority for the conservation of Earth's biological riches.

Gaoligong is an equally dynamic crossroads of culture and history. The valleys of the major north-south flowing rivers, the Nujiang and Longchuanjiang, have been farmed since ancient times. The Southern Silk Road, which crosses the southern portion of the mountain range, has connected India, Afghanistan, and Pakistan with central China since the 4th century B.C., serving as a conduit for commerce, trade, and culture. Today, approximately 450 families live in the eight hamlets that comprise Baihualing village, which is adjacent to the Gaoligong Mountain National Nature Reserve, the focus of our inventory (see figure 3). These villages reflect the remarkable cultural diversity of Yunnan, including Han, Bai, Lisu, Yi, Hui, and Dai ethnic cultures.

Although designated a Biosphere Reserve by UNESCO in October 2000, and designated a National Nature Reserve by the State Council of China, the spectacular environment of the Gaoligong range continues to suffer intense pressure. The unprotected lower slopes of the mountains contain great biological diversity, which is increasingly threatened as the forest cover rapidly disappears (figure 2C). The Chinese Government has selected a portion of the reserve for ecotourism development. Long-term survival of the Gaoligongshan natural treasures depends on full integration of resident communities in the management and stewardship of the reserve. Ecologically and culturally sensitive economic options—including carefully planned ecotourism—that directly benefit the region's human and biological communities are critical to protect Gaoligongshan's globally unique natural riches.

Overview of Results

ECOLOGICAL PROFILE

The Gaoligong mountain range in southwestern China is a dynamic blend of extremes—at once, an imposing geographical barrier and a major migratory passageway for plants and animals from east and west, north and south. With ancient peaks that rise abruptly to 4,000 m, Gaoligongshan is a hotbed of speciation. The lack of drought and minimal impact of glaciation offer conditions that allow primitive species to survive, especially in the low and mid elevations. Ecologically, Gaoligongshan is complex and unique: it is the one region on Earth where an intact and extensive transition from moist tropical to temperate forests still exists. In Gaoligongshan, the conditions that create new species and that prevent the extinction of old ones have persisted over a vast expanse of geography and time, in a region that is also home to ancient trading routes and a rich melding of cultures.

In the extreme southern portion of the range, the Gaoligong Mountain National Nature Reserve protects 405,549 hectares of forests along the border with Myanmar (Burma) (figures 2, 3). The reserve and its immediate surroundings were the target of our rapid biological and social inventories during the monsoon season of 2002, and of our focused analysis of physical structures to support ecologically and culturally sensitive ecotourism in the region. Surrounding, but not inside the reserve, live close to 300,000 people of 16 different ethnic groups: Han, Dai, Lisu, Hui, Bai, Miao, Yi, Zhuang, Nu, Achang, Jingpo, Wa, De'ang, Naxi, Drung, and Tibetan.

The Southern Silk Road (figures 8G, 9B), which traverses the Gaoligong mountains (figure 2), linked China, as far back as 300 BC, with India, Afghanistan and further west. The Road starts in Chengdu, Sichuan Province, crosses Yunnan, reaches the City of Baoshan and then scales the Gaoligong mountains, from Baihualing, before going into Myanmar. The Southern Silk Road was accessible year round because of the warm climate of the regions it traversed, and brought traders from as far away as Rome. Today, numerous historic sites survive along this ancient trade road in the Baihualing area, including an ancient stone arch bridge known as Huangxinshu, and the town of Jiujiez, a major trading post until the first highway bypassed it in 1958.

The Gaoligongshan region was a focus of activity during World War II, when Allied soldiers and local Chinese laborers built the Burma Road which made it possible to transfer supplies from British-controlled Burma into China. Landmarks of this era, when Chinese and Americans pushed back the Japanese invasion of China, include the Zhaigongfang battlefield at one of the major mountain passes between the eastern and western slope of the Gaoligong mountains.

Despite the steady and once-heavy traffic along the picturesque cobble Silk Road (figures 8F, 9B), the character of the vegetation it traverses is essentially intact above 2,000 m. The exception may be at Nan Zhaigongfang, the pass over the mountains at ca. 3,200 m (figure 3E), where the low, scrubby vegetation is 1,000 meters below the characteristic treeline. The scrub is the result of steady disturbance, grazing by pack animals, and collection of firewood by travelers over the millennia.

Below the 2,000 m limit of the reserve, local residents have exploited the forests heavily for timber and pasture over the years (figures 2B, 2C, 3D). Even the remaining wooded patches are degraded. Yet these pockets of forest still maintain a large fraction of the diversity found historically in the region. These lower elevations are rich in species, and contain several of the interesting and range-restricted species of plants and animals that we found during our inventory, including three species of plants new to science.

Gaoligongshan is largely a granitic range that has been heavily eroded. The vegetation on the eastern slopes is characteristic of the acidic (granite) substrates. On the west side, layers of volcanic ash provide a different, less acidic geochemical environment for the development of vegetation communities. We found major differences in species composition along the eastern and western slopes for all organisms we sampled (plants, fungi, and animals). The change in species with elevation also adds to the impressive richness of the reserve.

Gaoligongshan is cool and dry in the winter months (November—April), and warmer and much wetter during the summer (May—October). The forests cloaking the slopes are mainly monsoonal broadleaf evergreen forests. Patches of coniferous forest and scattered conifers within the broadleaf forest increase the diversity of habitats in the reserve. Abundant species of *Rhododendron* (figures 1, 4A), especially on the acidic soils of the eastern slopes, are a magnificent element of the Gaoligongshan understory during the flowering season. At high elevations, bamboos dominate the understory (figure 2F) and elsewhere bamboo species invade disturbed areas.

Streams cascade down the Gaoligong mountainsides, cutting deep valleys and creating majestic waterfalls (figure 2A). These fast-flowing streams (figure 2E) are crucial habitat for a number of animals, including several birds and amphibians restricted to the region. The valleys also provide habitat for different plant communities. Most valley-bottom plants rely on birds for seed dispersal, while plants that rely on mammals or wind to disperse their seeds dominate the ridges between the valleys.

Even during our brief sampling in three areas of the reserve, we added several new records to the existing species lists, highlighting the biological richness of the region, and how much more there is to be found in the reserve. In our brief social survey, the first of its kind in the Baihualing village gateway to the reserve, we confirmed the cultural richness of the local villages, along with the high potential for ecotourism and the need to strengthen local villages before tourism activities can be successful. Our visit also gave us the opportunity to develop preliminary designs for an ecotourism lodge and visitor center that can (1) share infrastructure with the existing ranger station, (2) be sensitive to the environment, and (3) be attractive to the ecotourist market.

In the following sections we summarize the highlights of our results and outline our recommendations for conservation action and for ecotourism development.

CONSERVATION HISTORY

In the first few decades of the 20th century, two of the world's best-known naturalists—Joseph Rock, who often worked for the National Geographic Society, and George Forrest, who worked for the Edinburgh Botanic Gardens—made important discoveries during their investigative trips to Gaoligongshan and the Baihualing area.

The Gaoligong Mountain National Nature Reserve was first designated by the provincial government in 1983 and the national government in 1986. In 2000, it became a UNESCO World Biosphere Reserve. Also in 2000, the boundaries of the National Reserve expanded significantly, and discussions are currently underway about further expansion. The following timeline provides additional detail.

1983: Provincial government designates a Reserve with boundaries including parts of the Baoshan District and the Nujiang Prefecture.

1986: State Council of China establishes the Gaoligong Mountain National Nature Reserve with somewhat different boundaries than the provincial reserve. The National Reserve includes 123,900 hectares (306,033 acres), and is 135 km (83.85 miles) long from north to south, and 9 km (5.59 miles) wide from east to west.

1994: State Council of China designates small portions of the reserve—8,550 hectares (21,119 acres) total—for tourism development. The tourism area includes three sections: Pianma (3,649 hectares) in the north, Rhododendron (270 hectares) in the west, and Baihualing (4,361 hectares) in the east. Baihualing is selected to be the first for tourism development.

2000: State Council of China triples the size of the reserve to 405,549 hectares (1,001,706 acres), largely by including the previously designated provincial reserve in the Gaoligongshan range (Nujiang prefecture) to the north.

2002: Discussions proceed on extending the boundaries of the reserve from the Nujiang River on the east to the Longchuanjiang River on the west (figure 3).

BAIHUALING VILLAGE: RESOURCES AND CAPACITIES

Contributors/Authors: Victoria C. Drake, Anne P. Underhill, Yuming Zhu, Ken Hao, Jianmei Yang

We conducted a social survey with the help of a local guide and an interpreter over 13 days. Baihualing village consists of eight hamlets: Hanlong, Upper Dayutang, Lower Dayutang, Bangwai-Guxingzhai, Taoyuan, Laomengzhai-Baihualing-Malishan, Manggang, and Manghuang (Appendix 6 Table 1). Each of these hamlets has its own leader. In each hamlet, we interviewed the leader and his family, and another local family. Our goal was to understand the community's capacity to support, and benefit from, tourism. Appendix 6 provides the list of questions along with detailed tables summarizing the inventory results.

The composition of the hamlets seems to have undergone many changes over the years. In some cases, single hamlets have split into multiple hamlets as a result of population growth. In other cases, small neighboring hamlets have expanded and combined to form one hamlet. Baihualing village has approximately 2,100 people and more than 450 families. Each hamlet has both wetlands and dry lands for farming. The leader of Baihualing village (Head of the Village Committee) is also the Communist Party Representative, with an office in Baihualing hamlet.

ETHNIC CULTURES

The major ethnic groups living today in the Baihualing area are the Han, Dai, Bai, Lisu, Yi and Hui (Muslim). Although these ethnic groups maintain separate cultural identities, cultural diffusion has occurred over time, accelerated in the last two decades with the end of the social stigma against intermarriage. We found tremendous flexibility in expression of individual ethnic identity. Individual children can decide whether to adopt the ethnicity of their father or mother, which often are different. Residence patterns also vary greatly, as the composition of multi-generational families often changes after a marriage. A man can move in with the family of his wife, if the wife's family has no sons and wishes to keep land within the family.

Most people choose to speak Chinese rather than their traditional language, although some people remember how to speak Lisu and Yi. Individuals may decide to retain their traditional language if there are perceived benefits. For example, some people learn Lisu to engage in economic transactions more effectively. Missionaries introduced the Lisu written language; the Bibles used in the Manggang church are in Lisu. Economic challenges—including farming and other ways of making a living in the rugged, mountainous terrain of Gaoligongshan—have led people to drop customs such as wearing of traditional clothing, preparation of specific food dishes, and production of traditional crafts. Most families celebrate Han festivals such as the Spring Festival or Chinese lunar New Year. But most families here also celebrate a traditional Yi festival, the Torch Festival. More in-depth research in the area will reveal additional customs and beliefs that distinguish the different ethnic groups.

NATURAL ASSETS AND CHALLENGES

Each hamlet faces the challenges of making a living with a distinct set of social assets (Appendix 6 Table 2). We estimate that the annual income per family for Baihualing village as a whole ranges from c. 300 to 20,000 Chinese dollars, with an average of c. 2,000 dollars. Farming is stressful for all families. People from every hamlet told us that they wish there were more means to invest in better technology and fertilizers. Low-lying hamlets have more water and wetlands available for rice. Most families grow some rice for their own subsistence, but depend heavily on sugarcane as the cash crop, from which they derive up to 85% of their income. The families' welfare is tied to the state-run factory nearby that buys the cane and also provides the seeds. Transportation of goods and people is difficult; the roads are steep and made worse by the frequent rains. Most people walk between hamlets; there are few cars, trucks, or other forms of transportation.

Fuel is a serious problem for everyone. There are few options beyond collecting dead wood and brush. Pressure for development of alternative sources such as methane is increasing. A factor that limits accessibility of natural fuel sources is ownership— or rights to the use—of land. The people in Hanlong, the hamlet closest to the Ranger Station and to tourist trails, pay the most attention to the protection of forests in the reserve. Some families here collect wild mushrooms and herbs to supplement their incomes (figures 8A, 8C), while others collect wild honey. Over 90% of families in Baihualing have adopted the new fuel-efficient stove, resulting in over 60% savings in energy. Apparently some families in Malishan continue to use the old "tiger stove" that wastes fuel. The development of tourism in the area will put even greater pressure on fuel resources.

Residents say that the Nujiang River is cleaner than it has been for years. However, heavy use of chemical fertilizers for farming and chemical pesticides for killing mice and rats in homes continue to affect water quality.

Basic expenses are a strain for everyone: fertilizers for farming, pesticides, education, health care, marriage, and funerals. People in each hamlet have devised creative methods of supplementing their family incomes (Appendix 6 Table 3). Perceptions vary about which hamlets have the most economic disadvantages, although more people in Laomengzhai and Malishan felt that their greatest need was for farming assistance. Some individuals recognize a connection between agricultural diversification, environmental protection, and financial well-being. Some families have shown that it can be very productive to grow a variety of fruits such as mango, lychee, and orange.

CULTURAL AND HISTORICAL ASSETS

Modern cultural and historical assets vary by hamlet (Appendix 6 Table 4). The earliest recorded human settlements in the area of the present Gaoligong Mountain National Nature Reserve date back at least

2500 years, to the Ailao people. The Ailao Kingdom, centered in Baoshan, flourished between 500 to 100 BC.

The earliest known physical traces of past peoples in the area come from the Ming period (1368-1644). An especially significant spot is a partially standing wall on a portion of the Southern Silk Road, near Hanlong. Another notable relic is a group of broken stone tablets bearing writing from the Ming period, at the site of the current Guangyin temple. According to local residents, these tablets come from an earlier temple dedicated to Guangyin. A study of records from the Ming period about the Gaoligongshan area should be made in the future. The museum established by Mr. Wu in Hanlong about the war against Japan is certain to interest visitors, although it will require additional preparation, especially because it contains unexploded bombs. Some individuals continue to make attractive hand-embroidered shoes (figure 8E) and aprons, baskets, gourd scoops, wicker stools, horsehair rain capes, and horsehair saddles.

SOCIAL ASSETS

Established groups in Baihualing village already work to improve economic conditions and promote conservation (Appendix 6 Table 5). The MacArthur Foundation funded the Gaoligongshan Farmers Biodiversity and Conservation Association (GFBCA) program in 1995. It was the first program of its kind in China. The purpose is to promote greater protection of the environment. More families would like to be chosen as demonstration or pioneer families so that they too can be involved in the process. Individuals in Baihualing have initiated other creative organizations, such as the tourism association in Upper Dayutang.

Education is a priority among people in most hamlets (Appendix 6 Table 6). Few people can attend and finish senior high school or university, since it is difficult to raise funds for the tuition. Family members are willing to make substantial sacrifices with respect to saving cash or replacing labor for farming so that a young person can attend school. After graduation, however, there are few options for young people to work outside the village, with the exception of a government job. Some people saw a connection between education, economic development, and conservation.

OPINIONS ABOUT TOURISM

Some people also recognize a relation between tourism, conservation, and economic development (Appendix 6 Table 7). However, very few villagers have experience with tourists, and there is little understanding of how tourism will affect them. People who are eager to provide services to tourists will need start-up funds to invest in such ventures. People indicated that they want to preserve the unique natural features of the area. Production of traditional crafts and food must take environmental conservation into account. People welcome training seminars about tourism and its relationship to conservation, and education will be a priority to help residents understand and benefit from tourism. Residents want to have a voice in the development of their area and are concerned about topics such as the location and management of potential markets for tourists (to sell foods and crafts), and continued protection of the reserve as tourism increases.

VEGETATION AND FLORA

Contributors/Authors: Jun Wen, Robin Foster, Jiali Qin, Shiliang Meng, Lilan Deng, Xiaochun Shi

The vegetation in southern Gaoligongshan is extremely complex. In addition to elevation, plant composition is directly related to history, especially colonization following major disturbance. Plant composition is also related to soil, especially with regard to the occurrence of mid-elevation elfin forests. The three major sites we inventoried—Baihualing, Datang, and Nankang (figure 3)—had remarkably different plant communities. We noted vegetation composition in relation to elevation, substrate (rock, soils), topography, and disturbance.

The changes we found in plant composition from low to high elevations and from east to west slopes are striking. The mountains act as a barrier to species dispersal between east and west slopes, although it may be possible for species to migrate around the southern end of the mountain range. Geological substrates also explain much of the difference between east and west: the eastern slopes of Gaoligongshan are acidic granite, while the western slopes have a deep volcanic sediment on top of the granite and are consequently more alkaline. As an example, the genus *Rhododendron* (figure 4A) has many more species and larger populations on the acidic eastern slopes. The genus *Acer* had different species on eastern and western slopes.

On a finer scale, isolated pockets of substrates increase Gaoligongshan's overall plant diversity by allowing different species to grow on each "island." For example, while species richness is low in the Ericaceae elfin forests covering ridges of quartz sand mixed with clay, the composition of species differs among the scattered, isolated islands of elfin forests. The trail to the waterfall at Baihualing (figure 2A) contains a dramatic example of the effects of substrate: 90% of the flora on the ridge slope before the waterfall is different from the previous ridge slope.

Along the gradient of elevation, changes in vegetation are mostly associated with frequency of cloud contact and relative lack of drought (figure 3A). On the mountain slopes above Baihualing there appeared to be two important altitudinal transitions in the vegetation: one near 2,000 m and another between 2,600 and 2,700 m. These represent average cloud contact during winter at the higher elevation and during summer at the lower elevation. Some mixing of plant communities occurs at the same elevation, depending on ridge exposure and ravine moisture. At the highest elevations and mountain crests, temperature, wind, and lightning become important factors shaping the vegetation. On exposed ridges, conifers and other emergent trees had a high frequency of lightning damage.

As expected, we found much higher species richness in the lower slopes than in the highlands.

Interestingly, these lower slopes are also where we found unusual or new records (see below). The lower elevations fall outside the protected zones in the Gaoligong Mountain National Nature Reserve, and are under severe and increasing pressure by surrounding human populations (figure 2C, see threats).

Landslides in Gaoligongshan appear to be relatively infrequent and small in scale, reducing what is often a major source of natural disturbance in other montane forests. Although also rare, fires do seem to have influenced forest dynamics, especially on the mountain crests and in elfin forests during years of extreme drought. Animal dispersal of seed has had a strong influence on plant composition. On the higher ridges and slopes we found primarily mammal- and wind-dispersed species, while on the lower slopes and valley bottoms we found plants that are primarily dispersed by birds. The more abundant species of plants, and the seasonality of their reproduction, have a major effect on the composition of the fauna and vice-versa.

Single species can dominate in areas besides the Ericaceous elfin forests. In mid-elevation ravines, we found extensive, dense, even-aged stands of *Strobilanthes* (Acanthaceae), a genus known to flower and die synchronously after growing for 10 to 20 years. We also observed many dense patches of bamboo, which has a similar ecology and colonizes landslides during fruiting years.

FLORISTIC RICHNESS AND COMPOSITION

Species richness and endemism are high in the flora of Gaoligongshan (figure 4). The botany team estimates a minimum of 5,000 species in the Gaoligong range, of which close to 4,500 have been recorded and described to date (210 families, 1,086 genera, and 4,303 species and varieties, Xue et al. 1995, Li et al. 2000). In Datang, we found at least three species that are new to science (two Araliaceae and one Vitaceae—see *Panthenocissus* sp. nov., figure 4G). About 10% of the flora—434 species—are known to occur only in Gaoligongshan. During our nine days in the field, we identified about 1,000 species of vascular plants (including six new records for the reserve), collected 300 species, and photographed more than 250 species. Most undescribed or unrecorded species are likely to be in poorly known plant groups, especially pteridophytes (ferns), Orchidaceae, Rosaceae, Ericaceae, Asclepiadaceae, Vitaceae, Urticaceae, Labiatae, Gesneriaceae, and Theaceae.

Ferns (pteridophytes) are very diverse in the reserve (see Appendix 2) and had not been previously surveyed systematically. Many of the species grow in mid-elevation evergreen forests. We collected 30 families (including five that are new records for Gaoligongshan: Equisetaceae, Adiantaceae, Drynariaceae, Loxogrammaceae, and Azollaceae) and 52 genera (8 new for the region: *Equisetum, Cheilanthus, Aleuritopteris, Adiantum, Woodwardia, Drynaria, Loxogramme,* and *Azolla*). The new records all came from Datang. Species richness in Pteridaceae, Dryopteridaceae, and Polypodiaceae are especially high, and we expect a large number of new species to be present in these complex families.

THE THREE REGIONS SAMPLED

Baihualing. On the eastern side of the Southern Silk Road, an almost complete elevational transect of plant communities stretches from the Nujiang (or Salween) River to the high mountains near Nan Zhaigongfang (figures 2, 3). In our transect along this road, from the Baihualing research station at 1,525 m to the crest of the mountains at 3,100 m, we recognized at least three major vegetation zones associated with elevation and/or topography. Each of the three vegetation zones had two or more highly distinct plant communities that were based on different geological substrates.

Near the hot spring close to the research station, monsoon evergreen broadleaf forests cover the lower foothills. Around Huangzhuhe (2,000-2,800 m), the first overnight camp structure along the Southern Silk Road, the mid-elevation wet evergreen forests are rich in epiphytes (Orchidaceae, Araliaceae, and Polypodiaceae). This area is highly diverse in Fagaceae, Lauraceae, Magnoliaceae, and Ericaceae. Dominant species include *Lithocarpus variolosus, Illicium simonsii, Juglans regia, Castanopsis* sp., *Cyclobalanopsis lamellosa, Elaeocarpus* spp., *Manglietia insignis, Acer* sp., *Michelia floribunda, Hydrangea* sp., *Fargesia edulis, Brassaiopsis palmipes* (figure 4F), *Brassaiopsis hainla, Merrilliopanax listeri, Symplocos* spp., *Vaccinium* spp., *Pteris* spp., and several species of *Rhododendron.* On the high mountain slopes around Nan Zhaigongfang, at 2,800-3,200 m—the camp at the pass (figure 3E)—the main species in the mountaintop shrublands are *Rosa* sp., *Rubus* sp., *Sorbus* sp., *Gentiana* sp., *Pedicularis* spp., *Rhododendron* spp., *Lithocarpus craibanus, Lithocarpus hancei, Fargesia* spp., *Mahonia polyodonta, Schefflera shweliensis, Daphne* sp., *Rumex nepalensis,* and *Caltha palustres.*

Datang. Datang, on the western slope of the Gaoligong Mountain National Nature Reserve (figure 3), has been less studied than Baihualing. Despite the severe disturbances it has undergone (from logging, cultivation and drying of tobacco, cattle, collection of firewood) Datang's flora is still very rich in species. We found an impressive number of unique or unusual plants, including three species new to science. Among the range-restricted species we recorded are the endangered *Rhododendron protistum* var. *giganteum, Taiwania flousiana,* and *Alsophila spinulosa,* as well as *Tetracentron sinense, Panax variabilis,* and *Aralia pausiloculata.* The dominant species we recorded are *Lithocarpus* spp. *Prunus nepalensis, Acer davidii, Daphniophyllum chartaceum, Pieris formosa, Evodia* sp., *Lindera communis, Decaisnea insignis, Tetracentron sinense, Manglietia insignis, Michelia* sp., *Edgeworthia gardneri, Embelia floribunda, Parthenocissus* spp., *Tetrastigma* spp., *Cladrastis sinensis, Ternstroemia gymmantheca, Akebia trifoliata, Oleandra* sp., *Rosa omeiensis, Pyrularia edulis, Iris tectorum,* and several species of *Elaeocarpus.*

Nankang. Nankang is at the pass, along the southern tip of the Gaoligong Mountain National Nature Reserve (figure 3). A relatively low-elevation pass for the region at 2,150 m, Nankang is an important biological corridor between the eastern and western slopes of Gaoligongshan. We found that Nankang shares many plant species with both Baihualing (east slope) and Datang (west slope). The mid-elevation wet evergreen forests in this foggy, high-rainfall site are rich in epiphytes such as orchids and ferns. Dominant species in the forest include *Lithocarpus variolosus, Schima khasiana, Phoebe* sp., *Castanopsis lamellosa, Michelia velutina, Cinnamomum caudiferum, Rhododendron delavayi, Rhododendron decorum, Schefflera elata, Manglietia insignis, Helicia shweliensis, Illicium macranthum,* and *Maianthemum purpurem.*

FUNGI

Contributors/Authors: Gregory M. Mueller, Bin Yang, Shangyi Ge

Gaoligongshan is extremely rich in macrofungi. Despite limited sampling, more than 300 species have been recorded in the reserve to date. During our nine days in the field, we registered over 200 species (of which we documented 148 with voucher specimens and color photographs; see Appendix 1, figure 7), even though we were able to inventory only a few of the vegetation types at a very limited elevational range. Only 22 species we found were among the 132 previously listed for the reserve. Although it is difficult to estimate the total number of macrofungal species — large ascomycetes, agaricales (mushrooms and relatives), aphyllophorales (bracket fungi, coral fungi, others) — that occur in the reserve, based on the diversity of plants, vegetation types, and altitudinal range, we estimate that the Gaoligong Mountain National Nature Reserve protects 1,500-2,000 species of macrofungi.

The remarkable mixing of species from different realms in Gaoligongshan underscores the global importance of the region for the conservation and further study of macrofungi. We observed two species of *Dictyopanus*, a tropical genus, in juxtaposition with a number of north temperate species that appear to be at the southern end of their distribution

(*Oudmansiella yunnansis, Lactarius lignyotus* and *Laccaria bicolor*). We found species of temperate Asia (*Amanita rubrovolvata,* figure 7D, and *Leccinum virens*) mixed in with species of temperate Europe (*Oudmansiella muscida*). And we recorded species that appear to occur in disjunct populations in eastern North America and eastern Asia (*Xerula furfuracea*). We also collected species endemic to China, like the large and showy *Boletus sinoaurantiacus* (figure 7E) that is restricted to the southern portion of the country.

Of the 148 species collected, we found only two in all three localities. We cannot determine the extent to which the lack of overlap reflects geography and site specificity, or limited sampling. However, each site clearly contains a unique assemblage of species. Distribution of fungi is strongly linked to vegetation, soil type, and available moisture among other variables. The geologically complex Gaoligongshan range, with its rich vegetation, should harbor many distinct macrofungal communities.

We sampled opportunistically during our brief period in the field, relying on local guides to take us to different sites known for their fungi. Fruiting of macrofungi is highly seasonable; sampling during different months of the rainy season would result in different records. We also know from experience that there is high year-to-year variation in fruiting of macrofungi, so we would find a number of different species if we re-sampled the same sites in subsequent years. We recommend much more extensive surveys. Although we were not able to collect quantitative data (which would have required three to five days at each subsite), we recommend that such samples be undertaken at least twice per year—at the beginning and end of the wet season—for several years. Such data will allow for comparisons in diversity and abundance among sites, within and outside of Gaoligongshan.

Macrofungi play a central role in forest and grassland communities. Preserving their diversity and populations is critical for the maintenance of plant and animal communities. Macrofungi play vital roles in nutrient cycling and absorption, water regulations, plant-to-plant interactions, and interactions among other soil-associated organisms. Many macrofungi form a symbiosis with Fagaceae, Pinaceae, and other trees, dominant elements of the canopy in Gaoligongshan's forests. These ectomycorrhizal associations are essential and mutually beneficial to both tree and fungus. While trees provide sugar to the fungi, the fungi provide the trees with minerals and water and protect them from root pathogens. The fungi can even link trees together below ground, enabling trees to share carbohydrates and minerals and creating a dynamic, interacting community. Other macrofungi are crucial decomposers: macrofungi, along with some bacteria, are the only organisms capable of decomposing cellulose and lignin, the two primary constituents of plant material. These fungi are the primary recyclers in plant communities. A few macrofungi are pathogens of plants, and promote important age diversity in forests.

Wise management and use of macrofungi can become an important component of the ecologically compatible economic alternatives for Gaoligongshan. People in the region use macrofungi for food and medicine; these fungi constitute an important component of the local diet and also supplement family income. The beauty of mushrooms and other macrofungi (figure 7) will add significantly to the experience of ecotourists, as will the delicious (and for foreign visitors, exotic) mushroom dishes.

Fungi can provide a crucial link between human communities and biological communities in Gaoligongshan. Studies in the northwestern United States and southwestern Canada have documented that the economic value of non-timber crops (primarily mushrooms, ferns, and greenery for the florist market) can exceed the economic value of trees as timber. Fungi, thriving in high-quality well-managed natural communities, may become a valuable, renewable resource directly benefiting neighboring human communities.

AMPHIBIANS AND REPTILES

Contributors/Authors: H. Bradley Shaffer, Yu Zhang, Tianchan Wang

The herpetological team registered 30 species of amphibians and reptiles during less than seven days of sampling (Appendix 3). We found seven snakes, four lizards, one salamander, and 18 frogs (with possibly 15 to 21 frogs; determination awaits more careful studies in collections; figures 5 and 6). Even with our limited sampling efforts, we found at least five species not reported among the 72 previously registered for Gaoligongshan (Xue 1995). These are the snakes *Pytas mucosas, Rhabdophis nuchalis* (figure 6C), and *Rhabdophis subminiatus,* and the frogs *Chrixalus doriae* and *Micrixalus liui.* With 15% of our records new for the region, we expect that many more unrecorded species exist in the Gaoligong Mountain National Nature Reserve. Based on species known for Gaoligongshan and species registered in Yunnan and in the southwestern mountains of Yunnan, we expect some 60 reptiles (with a predominance of snakes) and 60 amphibians (with a predominance of frogs) to occur in the reserve. This is a high number even relative to the species-rich Amazonian tropics. Because of their often-cryptic habits, we believe that this undiscovered species richness will be most extreme for snakes and lizards. We probably only scratched the surface of the true diversity of snakes in the reserve. We also suspect that many of the supposedly wide-ranging species of frogs contain cryptic species with more restricted distributions.

We conducted night surveys for frogs based on their calling at streams, ponds, pools, and rice fields. Generally, three individuals worked at different stretches of the habitat, and we attempted to locate all species that were calling. We also conducted visual surveys for non-calling individuals using flashlights and headlamps. During the day, we conducted one or two visual surveys by walking along or adjacent to trails, turning logs and rocks, and searching in trees and bushes for lizards and snakes. We also walked along streams and pools in search of adult and tadpoles of frogs and toads.

The Gaoligong mountains represent an important mixture of mid-elevation subtropical species (1,500-2,400 m) and high-elevation species (above about 2,700 m). The high-elevation species of amphibians are the most geographically restricted, with frogs of the family Pelobatidae showing the smallest geographic ranges. For example, the frog *Scutiger gongshanenis* is restricted to the Gaoligongshan region above 2,500 m, where we found it in clear, fast-flowing headwater streams. The frogs *Leptobrachium chapaense, Rhacophorus gongshanensis* and the lizard *Japalura yunnanensis* are also restricted to the region, but we did not find them during our short visit. We expect that several of these mid-to-high elevation taxa are actually composed of several distinct species, and that more intensive field collections will reveal additional species that occur only in Gaoligongshan.

At approximately 2,700 m, the composition of amphibian and reptile communities shifts abruptly from lowland to montane species. We heard *Megophrys minor* (figure 5D) at least up to 2,500 m, and probably up to about 2,700 m, at which point we ceased to hear its call. At 3,100 m at the top of the Southern Silk Road, we heard the distinctive call of *Scutiger gongshanensis* in the types of small headwater streams where we heard *M. minor* calling at lower elevations. We heard at least one other species of frog (whose identity we were not able to confirm) calling at 3,100 m that we did not hear at lower elevations. The habitat shift that we observed at 3,100 m, both in terms of climate and forest structure, made this elevation unsuitable to the mid-elevation species, such as the large snakes *Zaocys nigromarginatus* and *Pytas mucosas*.

Although our fieldwork was brief, our sampling confirmed several impressions. First, the Gaoligong mountains themselves appear to be a major biogeographic barrier to dispersal of several species of amphibians. In addition to our samples of amphibians, we were able to census populations by calls during our wet-season survey. For example, at the Baihualing Station on the Southern Silk Road (eastern slopes), the frog *Megophrys minor* (figure 5D) called from virtually every stream narrower than about 1 m, from 1,400-2,500 m. This was the case during rainy days or at night. However, we never heard this species in similar small streams in Datang, on the west side. If the species is present on the west slope, it is clearly much less abundant than on the east side, and our guess is that it does not occur at all on the western slopes. Similarly, the frog *Rana pleuraden* (figure 5F) was extremely abundant at Datang and Nankang but appeared to be absent at Baihualing. Finally, the salamander *Tylototriton verrucosus* (figure 5A) was abundant in the east (Baihualing; including in exceedingly disturbed habitats in the village), was present at Nankang (at a pond site), but was absent from appropriate habitats at Datang. Because these species were abundant when found and occurred in both undisturbed and disturbed habitats, we feel confident that if they had been present at all three sites we would have discovered them.

We believe that the mountains also serve as a major biogeographic barrier for reptiles. Our reptile sampling was not as complete as for amphibians, however, so more fieldwork is needed to confirm this impression. We found a pattern of patchy distribution for many of the reptiles similar to the pattern observed for frogs, but the low numbers of individuals we found makes us less certain that absences in our samples represent true absences at a site.

Clearly, to understand the extent of these community shifts will require much more extensive surveys, particularly for the reptile species. Virtually the only way to sample lizard and snake species effectively is to use drift-fences with associated traps. These fences are simple to install and census, but they require that a trained person, either from the reserve or from the local community, check the traps, identify individuals captured, and release the animals. Monitoring each set of drift-fences would require 1-2 hours per day, and they need to be checked daily. Fortunately, when it is inconvenient to check the traps, the drift-fences can be closed down to allow animals to pass by them freely. Reopening the traps when personnel are available requires only a few minutes.

Several members of our expedition noted that amphibians have declined in some areas around Gaoligongshan, particularly in the lowland agricultural areas. This observation is consistent with the worldwide trend in amphibian declines. One of the key reasons for these dramatic declines is the application of pesticides and herbicides, which kill amphibian adults, tadpoles, and eggs. Gaoligongshan has wetland ponds, pools and streams that are not subject to these agrochemicals, and the resulting high densities of common amphibians within the reserve is striking compared to adjacent agricultural regions. For example, in Datang, we found the same set of species in the rice fields (figure 3D) as we did at our field camp (about 5 km away) and at our sampling site at Nankang. However, in the rice fields species numbers were very low, with only 1-5 individuals per species found in two hours of intensive nighttime collecting. At the pools near our Datang field camp, we found 2-5 times as many individuals in the same time, and at Nankang, in the undisturbed botanical garden, we found 5-10 times as many individuals. The Gaoligong Mountain National Nature Reserve, protected from the application of deadly agrochemicals, is extremely important as a safe haven for frogs.

Gaoligongshan is a superb reserve for the amphibians and reptiles that live at mid-to-high elevations. At lower elevations, we believe that most species still exist, but in reduced numbers because of intense human activities. Species that inhabit the lower slopes appear to be recoverable if human activities are reduced, or if people change the ways that they interact with the landscape. Important changes include: reducing the use of agricultural chemicals; ending the practice of killing large snakes, lizards, frogs, turtles and tortoises; ceasing the clearing of trees particularly near any streams or rivers; and reducing the extent of cattle and goat grazing. Convincing people not to kill snakes can be particularly challenging, and might benefit from an educational element on the value of snakes for pest control.

BIRDS

Contributors/Authors: Douglas F. Stotz, Ruichang Quan, Zhengbo Li, Debra K. Moskovits

We estimate that the Gaoligongshan region supports over 600 species of birds—nearly one-half of all species known in China. The bird team registered 179 species in the Gaoligong Mountain National Nature Reserve (Appendix 4) during the nine days in the field: 121 in Baihualing (east slope), 104 in Datang (west slope), and 54 in Nankang (pass, at south). Approximately 350 species had been recorded previously in the reserve (Xue 1995), of which many are lowland species (below 1,500 m), waterbirds, or open-habitat species not expected to occur in significant numbers within the reserve. We added 23 species to the list, 18 of them forest species. Three of the new records represent significant range extensions: *Clamator jacobinus* (registered at the Southern Silk Road) represents a range extension to the east, previously known in China only from southern Xizang; *Enicurus immaculatus* (found in Datang and Baihualing) is also a range extension to the east, previously known in China from a single record at Tengchong; and *Tickellia hodgsoni* (found at Baihualing) fills a gap in range between Xizang and southeastern Yunnan.

The reserve supports an impressive montane avifauna (above 1,500 m), which we estimate at 355 forest species. Close to 225 forest-based breeders already have been recorded. Based on ranges of thus far unrecorded species, we expect that an additional 100 species breed in the Gaoligong forests. We estimate that 30 additional land bird species winter in or migrate through these forests (16 already have been recorded). In the surrounding lowland and disturbed areas, at least 130 additional species have been recorded and we estimate that future surveys will find at least twice that number.

The lower slopes of Gaoligongshan, below about 2,400 m, have a particularly rich array of species, many with small geographic ranges. At the three sites visited, we observed substantial differences in the avifaunas at equivalent elevations, even in the very common, easily observed elements of the avifaunas.

During our inventory, we were able to evaluate turnover of species with elevation only along the forested stretch of the Southern Silk Road (Baihualing Station to the pass at 3,100 m, figure 3). Each of the three subsites along the Southern Silk Road (1,500 m-2000 m; 2,000 m-2,800 m; 2,800 m-3,100 m) had distinct avifaunas and no species was found across the entire elevational range (although with further study, undoubtedly some species will range that widely). In a few genera, for example *Alcippe, Yuhina,* and *Aethopyga,* we found altitudinal replacement of species. However, this replacement was not as sharp as is regularly the case in tropical montane avifaunas. In general, elevational ranges seemed broader in subtropical to temperate Gaoligongshan than in avifaunas of the tropical Americas or Africa. And populations at some elevations may be maintained through immigration from other elevations. Specifically, the richness of species at about 2,000 m may be maintained by movements from lower elevations. If the forests at the lower elevations (see figures 2B, 2C, 3D), which are outside the reserve and face much pressure, are lost, much of the diversity higher in the mountains may be lost as well.

Almost one-quarter of the forest birds registered at Gaoligongshan have restricted ranges and therefore are likely to be at risk. Gaoligongshan has a number of very narrowly endemic species (range smaller than 50,000 sq km) representing two Endemic Bird Areas (EBA) as defined by Birdlife International (Yunnan Mountains and Eastern Himalayas, Stattersfield et al. 1998). We recorded five such species during our nine days in the field: two of the three narrowly endemic birds of the Yunnan Mountain EBA (*Sitta yunnanensis* and *Paradoxornis brunneus*; both at Datang), and three of the 22 narrowly endemic birds of the Eastern Himalayas EBA (*Tickellia hodgsoni, Sphenocichla humei,* and *Heterophasia pulchella*). The first three of these five species are new for the Gaoligongshan list. Three additional species from the

Himalayan EBA—*Lophophorus sclateri, Harpactes wardi,* and *Actinodera waldeni*—have been recorded from Gaoligongshan, but were not observed by us. In total, eight narrowly endemic species have been recorded from the Gaoligong range. We recorded another 37 species (22% of our total) with ranges in the order of 100,000 sq km or less.

Birdlife International has applied IUCN (International Union for Conservation of Nature) criteria to rate the degree of threat to all birds of the world (Collar et al. 1994). Based on this ranking we found 10 species that are threatened or near threatened with extinction. An additional nine species in these categories were recorded previously from Gaoligongshan. Habitat loss is the major threat for these birds. Seven species of pheasants, including five considered at risk by Birdlife, are a particularly important set of vulnerable birds. Current hunting pressure seems to be low, but the pheasant populations should be monitored. We must act now to ensure that those species that are "near threatened" do not become threatened, and to keep those that are threatened from becoming extinct. Gaoligongshan is tremendously important as a refuge, and it must be sufficiently large to maintain bird populations in the long-term.

Gaoligongshan is a rich biogeographic cross-roads, as indicated not only by the presence of species representing two endemic bird areas, but also by the mixture of elements from the avifaunas of the Himalayas, central China highlands, southeast Asia, and the Palearctic. Gaoligongshan is also near the center of diversity for the babbler family (Timaliidae, the taxonomy is currently in flux), which is mostly an Asian radiation. The reserve has 60 species of babblers, of which we saw 43. The babblers, a very diverse group ecologically, range across almost the entire array of shapes, sizes, and bill morphologies. Large genera at Gaoligongshan include the laughing-thrushes (*Garrulax,* 12 species recorded), fulvettas (*Alcippe,* 7 species), and yuhinas (*Yuhina,* 7 species).

We cannot yet evaluate the effect of seasonality on Gaoligongshan's avifauna. We observed a number of species with dependent young (Appendix 4), suggesting that we surveyed at the end of the breeding season for most species of forest passerines. Gaoligongshan has significant numbers of species that only winter in the region (about 35 species). It also has many species that breed in the area but winter elsewhere (about 45 species). In addition, a number of the birds at Gaoligongshan make seasonal elevational movements. These migratory patterns mean that successful conservation of Gaoligongshan's avifauna depends on careful management not only of the reserve but also of its surrounding areas.

LARGE MAMMALS

Contributors/Authors: Huaisen Ai, Rutao Lin

We found direct or indirect evidence of 42 species of large mammals in Gaoligongshan during our survey (see Appendix 5): 34 along the Southern Silk Road at Baihualing, 28 in Datang, and 21 in Nankang. We estimate that the Gaoligong Mountain National Nature Reserve supports about 150 species of mammals. The fauna is complex, with dominant elements from the Oriental region but with several species from the Palaearctic. The reserve is rich in endemic species (about 40) and in forest species (over 85% of fauna).

Our survey relied on direct evidence of animals through visual sightings or sounds, and indirect evidence through tracks, droppings and semi-structured interviews with villagers, especially hunters. To ensure reliable information in our interviews, we ran several trials with photographs of animals that do not occur in the region.

One of our most significant findings was that the lesser panda (*Ailurus fulgens*) occurs in areas around 2,000 m in Gaoligongshan. This species was believed to be restricted to alpine areas, above 3,000 m. We also found indications that the population sizes of the endangered *Hylobates hoolock* and *Cervus unicolor* are increasing and that their distribution within the reserve is increasing as well.

Gaoligongshan protects several rare and endangered mammals. In all, 27 species known from the Reserve are listed as first- or second-grade nationally protected species. We found tracks or individuals of 13 of them: *Macaca mulatta, Semnopithecus phayrei, Hylobates hoolock, Macaca arctoides, Selenarctos thibetanus, Ailurus fulgens, Catopuma temmincki, Cervus unicolor, Budorcas taxicolor, Capricornis sumatraensis, Naemorhedus caudatus, Ratufa bicolor.*

As with the other taxa, turnover of species with elevation was most evident in Baihualing, where there is a continuous stretch of forest from 1,500 m to the crest. Although Datang has large patches of forest, human disturbance is ever-present and severe in several places, especially in the lower elevations. Nankang also has had great impact from human activities that have essentially eliminated the understory. As expected, the smaller-sized large mammals (primates, muntjacs, hares) are more abundant than the larger ones (pangolins, black bears, and large cats).

THREATS

The Gaoligong Mountain National Nature Reserve protects a spectacular array of natural communities in the southern portion of Gaoligongshan. Through proper planning, the reserve can also offer concrete opportunities for integrated management with the culturally rich human communities living at the base of the mountains. Nonetheless, pressures inside and outside the reserve, if unmitigated, threaten the stability and long-term survival of the reserve and of its neighboring human and biological communities. Below we list these primary threats to the region.

1. *Continued degradation or complete loss of lower elevation habitats.* The lower slopes and valleys adjacent to the reserve are unprotected and are under extreme threat (the reserve reaches its lower limits at ca. 1,500 m). Only pockets of forest remain, most of them already degraded (figures 2B, 2C, 3D). Yet even in these degraded and increasingly isolated

pockets we found a surprisingly high number of rare, range-restricted, and/or endangered species. These lower-slope forests provide crucial habitat for species that do not occur elsewhere, and they help maintain populations of animals at higher elevations. All three plants new to science in our inventory came from the lower elevations at Datang.

The degradation or outright destruction of the lower slopes and valleys comes primarily through (i) clearing for agriculture (with cultivation and drying of tobacco being one of the most damaging practices); (ii) production of charcoal; (iii) conversion to pasture for livestock; (iv) grazing by livestock (also a threat at higher elevations, inside protected areas); (v) use of trees for firewood and for construction; (vi) erosion; and (vii) use of agricultural chemicals (which is even more devastating to streams, wetlands, and ponds).

2. *Disruption of streams and rivers and destruction of sensitive watersheds.* A number of animal species—including several that are restricted to Gaoligongshan—depend directly on streams for their survival. An even greater number of species rely indirectly on these streams. In the agricultural areas around Gaoligongshan, streams have been all but destroyed (primarily by diversion for agriculture and chemical pollution). Destruction of sensitive watersheds is particularly dangerous for very small headwater streams, which harbor a large portion of Gaoligongshan's amphibian diversity. Pollution of wetlands also affects the macrofungi, which in turn affects the plants and ultimately the structure of the vegetation and the availability of food for animals.

3. *Incursion of livestock in protected forests.* Cattle, pigs, and goats lead to fast deterioration of the forests. They cause direct damage to the vegetation, and also compact the soil, making it impossible for fungi and other organisms to live. The understory, and all creatures associated with it, will be the first to disappear if pressure from livestock remains high.

4. *Drift of pollutants (agricultural chemicals, others) into the reserve (see 2).*

5. *Overuse of resources: hunting, harvesting.* Over-harvesting of mushrooms may be or may become a problem, if not studied and managed so that sufficient fruiting bodies remain uncollected each season. Similarly, excessive hunting could become a problem for populations of large mammals and large birds, especially pheasants. While hunting and harvesting are controlled inside the reserve, and to a lesser extent in the collective land outside the reserve, there are areas that are not managed at all. All hunting and harvesting should be managed and the impact monitored.

6. *Pressure to generate income.* The pressure to develop income-generating activities inside or around the reserve—e.g., ecotourism—can open up excellent opportunities for collaboration with the neighboring human communities in ecologically compatible use of resources (see recommendations, below). However, unless managed carefully and adjusted based on a rigorous monitoring plan, such revenue-building initiatives may damage both natural and human communities through overuse, lack of infrastructure, increased pollution, and disturbance.

7. *Lack of information about and access to "green technologies."* The Farmers Biodiversity and Conservation Association has made great strides, but most residents in the villages surrounding the reserve lack basic information about environmentally safe farming practices and fuel sources. Issues about education, health care, and sanitation are dominant in peoples' minds and must be considered when approaching villagers about long- and short-term goals for conservation.

CONSERVATION TARGETS

The following species and communities are of primary focus for conservation in Gaoligong Mountain National Nature Reserve because of their (1) global or regional rarity, (2) influence on community dynamics, and/or (3) importance in ecosystem processes.

Organism Group	Conservation Targets
Biological communities	Low-elevation forests (below 2,000 m), even if disturbed
	Elfin forests and other communities on unusual geological substrates
	All intact forests and full elevational gradients (Southern Silk Road)
	All distinct vegetation communities on the eastern and western slopes
	Vegetation communities with different age structure and ample wood debris (to maintain fungi diversity)
Plants	Endemic taxa (plant species restricted to Gaoligongshan or with restricted ranges)
	Endangered species (*Rhododendron protistum* var. *giganteum, Taiwania flousiana, Alsophila spinulosa*, others)
	Plants used for medicine
Fungi	Fungi used for food and/or medicine
	Mycorrhizal fungi associated with each vegetation community and with particular plants (e.g., Fagaceae, Pinaceae)
	Decomposing fungi
Reptiles and Amphibians	High-elevation stream amphibians (especially above 3,000 m)
	Mid-elevation subtropical species (1,500 m—2,700 m)
	Species endemic to Gaoligongshan or ones with restricted geographic ranges
	Economically important species (e.g., monitor lizards, python, turtles and tortoises)
	Threatened low-elevation species
Birds	Montane evergreen broadleaf forest birds (especially below 2,000 m)
	Birds along (or using) fast-flowing streams
	Coniferous forest birds
	Large forest pheasants—seven species known from Gaoligongshan (three vulnerable and two near-threatened)
	Endemic taxa (especially narrowly endemic species)
	Threatened and near-threatened species

CONSERVATION TARGETS

Mammals	Primates: *Macaca arctoides* (stump-tailed macaque), *M. mulatta* (rhesus monkey), *Semnopithecus phayrei* (brow-ridged langur), *Hylobates hoolock* (white-browed gibbon)
	Large carnivores: *Selenarctos thibetanus* (black bear), *Ailurus fulgens* (lesser panda), *Panthera pardus* (leopard), *Panthera tigris* (tiger)
	Large ungulates: *Cervus unicolor* (sambar), *Budorcas taxicor* (takin), *Capricornis sumatraensis* (serow), *Naemorhedus caudatus* (goral)
	Rare and endangered species

Gaoligong Mountain National Nature Reserve offers safe haven to a spectacular array of biological communities and to a large number of range-restricted, and/or endangered species. In just nine days in the field, we significantly expanded the list of organisms known to exist in the reserve, underscoring (1) the importance of the region for the long-term survival of global biological treasures that occur nowhere else, and (2) how much more there is to learn.

The reserve offers the opportunity to conserve an unparalleled mixture of ecological communities, while integrating the diverse cultures at the base of the mountains in the management and restoration of the highlands and of the surrounding, unprotected lower slopes and valleys. The reserve can become an international model for ecotourism that brings economic benefit to local people, instills local pride, fosters an ethic of stewardship, and preserves indigenous cultures. Below we highlight long-term benefits that conservation and integrated management of the reserve will bring to the region and to the world.

1) **A globally important nature reserve—from the lower slopes at 1,500 m up to the rugged crests at 4,000 m—protecting biological communities from the Oriental realm of China, the Himalayas, and the Palearctic.** The continuous stretch of forest from low to high elevation and from east to west over mountain crests conserves a blend of biological realms unique in the world and protects the processes that bring about speciation.

2) **Restored habitat for an enormous diversity of plants and animals currently at risk on the lower slopes.** Pockets of forest still standing in the lower elevations outside the reserve's present boundaries are highly diverse, with unusual and/or range-restricted species of plants and animals. Protection and expansion of these pockets of forest, along with appropriate management efforts, will restore crucial habitat for a currently neglected, large segment of the plant and animal diversity in Gaoligongshan.

3) **A model for successful ecotourism that is ecologically and culturally sensitive, and that benefits both the local communities and the nature reserve.** Adjacent to the nature reserve are communities that reflect the exceptional cultural diversity of Yunnan Province (figure 8). Visitors and residents have a unique opportunity to experience the diversity of both nature and culture. Carefully planned tourism, with full participation from the neighboring villages, has the potential to attract funds for long-term management of the reserve and to integrate local villagers in an ecologically sensitive economy.

4) **Successful collaboration among neighboring communities and with reserve personnel in the management and use of the nature reserve.** Local citizens can become excellent stewards of the reserve. Appropriate training programs in Baihualing can serve three complementary purposes: enhance the quality of life of neighboring villagers, reduce stress on the natural environment, and provide the necessary social foundation to handle increased tourism in the area.

5) **Protection of watersheds and of other resources used for food and medicine.** Effective conservation in the reserve and restoration of the lowlands provide direct and immediate benefits to the surrounding communities and to the economy of the region. Watershed protection in the mountains is crucial for the continued supply of water to agriculture in the valleys. Local residents harvest several natural resources for food, medicine and additional income. Fungi, especially, make up an important component of the local diet and also are sold to supplement income. Managed use of macrofungi can be an important component of sustainable harvesting in Gaoligongshan.

6) **Center for studies of evolution (speciation), ecology (migration, habitat use, recruitment and growth of dominant tree species), and conservation (amphibian decline, effects of pollution).** The continuous swath of forest on the eastern slopes of the Gaoligong Mountain National Nature Reserve, and the isolated patches of forest on the western slopes just outside the reserve, are excellent sites to study conservation and biodiversity issues such as speciation, habitat use, and causes of amphibian decline.

The Gaoligong Mountain National Nature Reserve offers vital protection to the montane flora and fauna of southern Gaoligongshan, primarily from 2,000 m and above. Yet to realize its potential as a worldwide resource of unique biological communities, we envision an expanded reserve that also protects the species-rich, lower montane slopes. In this vision, the buffer zone for the expanded reserve (1) stretches from the Nujiang River to the Longchuanjiang River (figure 3) and (2) connects the reconfigured reserve to protected areas to the north. Meanwhile, ecologically compatible economic activities, including sensitive ecotourism, thrive under integrated management activities that strengthen existing community assets and support the villages at the base of the mountains.

The following recommendations propose crucial steps toward realizing this vision. An active research and inventory program will sharpen and focus these conservation goals. In Appendix 7 we present detailed recommendations for development of an ecotourism lodge.

Protection and management	1) **Keep the core of the reserve untouchable, with a few areas open to researchers.** Because of the tremendous biological importance of Gaoligongshan's forests and the strong human pressure all around the reserve, we recommend that a significant portion of the reserve (the current "core area," which includes the higher elevations in the reserve, see figure 3) remain completely off-limits, as it is now.
	2) **Increase protection and strict patrol of the non-core areas of the reserve.** This increased protection in the lower elevations, where tourism also is allowed, will be crucial to maintain integrity of the forest.
	3) **Extend the core area of the reserve to lower elevations wherever possible.** There is an enormous diversity of plants and animals in the lower elevations of the reserve, including vulnerable species with restricted ranges. This diversity is at high risk of extinction because of continued invasion of the reserve and erosion of the isolated pockets of remaining forest outside the reserve (figure 2C).
	4) **Extend conservation-compatible management from river to river.** Currently, discussions are underway about extending conservation-compatible management from the Nujiang River on the east to the Longchuanjiang River on the west. While preservation of the resources within the river valleys is critical, an expanded designation of protected status must take into account the approximately 300,000 residents of the river valleys. It may be appropriate to consider a "heritage area" designation outside the present reserve boundaries that allows for continued use of the living landscape with appropriate incentives to encourage cultural preservation and ecological restoration, and restrictions to limit development to ecologically sensitive sustainable activities. As an immediate next step, beyond the official boundaries of the reserve, collaborative

programs should be developed with neighboring villages to manage the natural resources outside the reserve in a manner that is compatible with the maintenance of biological and cultural diversity.

5) **Restore and protect the patches of low-elevation forests surrounding the reserve; extend currently isolated forest patches eventually to link one to the other and to the larger protected areas.** For the long-term survival of global biological treasures in the low-elevation forests—where we found species new to science during this brief inventory—we recommend working closely with the neighboring villages to develop and implement collaborative programs that (i) restore remaining patches and (ii) reforest denuded stretches among patches with native species to increase available habitat. We recommend that all remaining patches of forests outside the reserve—even the small ones in Nankang and the highly disturbed patches in Datang—be afforded protection.

6) **Strengthen Baihualing village's capacity to participate in the planning and implementation of ecotourism.** One opportunity to ensure that local residents are involved in planning and benefit from tourism activities would be to establish a vigorous village ecotourism association (in the fashion of the Farmers' Biodiversity and Conservation Association) representing the eight hamlets. This association would work with the existing village committee to discuss and implement plans and policies related to tourism. To be successful, this tourism association would function at a larger scale than the existing small committee in Dayutang, and would represent all hamlets.

7) **Research and implement ecologically compatible agricultural practices in the lower slopes and valleys and seek opportunities for ecological restoration.** Increase options for farmers to diversify crops and to reduce use of polluting fertilizers and pesticides (which are also extremely expensive) and to benefit economically from ecological restoration activities.

8) **Increase affordable options for fuel.** Currently wood is the primary source of fuel. Tourism will increase pressure on the forests by increasing demand for fuel. Local villagers are currently unable to afford alternatives such as methane.

Ecotourism

1) **Ensure that all ecotourism activities directly benefit the reserve and the neighboring villages.** The purpose of ecotourism, as well as of other ecologically compatible economic activities, is to bring revenues to the region that are consistent with the conservation of biodiversity and of the cultural and social assets of the neighboring villages. Tourism development should foster continuity of local cultural traditions, while recognizing and respecting the cultural change that continues to occur over time, and respecting the rights and privacy of local citizens.

   Specific recommendations include (1) use of local guides, (2) use of local foods, (3) a substantial entry fee with revenues applied directly to protection and maintenance of the reserve and to strengthening of the communities, and (4) an ecologically sensitive infrastructure.

2) **Research carrying-capacity for visitors in the reserve and manage visitor loads accordingly.** The Gateway Lodge and Visitor Center (Appendix 8) is designed based on the assumption that 100 visitors is an appropriate maximum to protect the ecology of the reserve. Further research is needed to verify the appropriate maximum number of visitors entering the reserve and its environs at Baihualing as well as other future reserve gateways.

3) **Design all ecotourism activities and infrastructure to minimize impact on the sensitive biological communities and to respect and preserve local cultures.** Unplanned development will put at risk the natural and cultural resources that give the area its internationally unique character and picturesque landscape. To protect the biological riches of the reserve, the number and activities of tourists must be managed tightly (especially on the Silk Road), with a program designed to monitor impact and to fine-tune the activities and infrastructure accordingly. Local citizens should participate in making decisions about how visitor activities are managed, to ensure protection of privacy and quality of life.

   a. **Create a Gateway Lodge and Visitor Center that minimizes consumption of limited natural resources, minimizes disruption to the reserve, is architecturally compatible with local design, builds on existing infrastructure including the Ranger Station, and keeps new infrastructure outside the reserve boundaries** (Appendix 8).

   b. **Regulate the number of visitors and strictly control visitor activities** (Appendix 8).

c. **Limit and monitor activities that can damage biological communities.**

   i. **Limit overnight trips within the reserve.** These trips cause the highest impact, given the need for infrastructure and cooking inside the reserve, and the use of pack animals.

   ii. **Limit activities within the reserve to well-maintained trails.** Trails should be built, used, and maintained to reduce erosion while preserving historical integrity (e.g., Southern Silk Road).

   iii. **Properly dispose of waste.**

   iv. **Prohibit or severely limit use of firewood for cooking and heating.**

   v. **Limit the use of pack animals.**

4) **Approach the Gaoligong Mountain National Nature Reserve as one of a constellation of tourism destinations within this part of Yunnan Province.** This will reduce pressure on the reserve and will create a rich visitor experience, while strengthening communities and preserving indigenous cultures and landscapes.

a. **Develop regional ecotourism plans and interpretation techniques that promote a high-quality visitor experience from the moment visitors enter the area at the Baoshan and/or Luxi airport.**

b. **Create a series of day trips to other regional attractions.**

c. **Develop an interpretive plan for the reserve in conjunction with an analysis of sensitive areas that identifies appropriate access points to draw visitors to different parts of the reserve.** Identify the scenic and natural qualities that would draw visitors to each area and would provide a series of different educational experiences.

d. **Develop a series of field guides and audio compact discs.** Audio CDs for birds and frogs would greatly enhance the visitor experience. Photo guides or short booklets should be developed for plants, mushrooms, amphibians and reptiles, and for surrounding historic sites.

  e. **Offer targeted but low-impact wildlife viewing experiences for heightened enjoyment.** For example, in areas of low sensitivity, create small (0.1–0.2 hectare), shallow ponds with seating and shelter where visitors can enjoy frogs calls at night, observe frogs and newts mating, and watch other wildlife attracted to water. (The pond at Nanking is an excellent example.)

  f. **Support development of amenities that serve local residents and tourists and that strengthen the social, cultural and historical assets of Baihualing.**

5) **Prepare residents of Baihualing village to participate effectively in and benefit from tourism development.** In addition to the measures to strengthen and build the capacity of the community, residents seek training and support to help them participate in a tourism economy. For example, classes in English and Japanese language, guiding/interpretation techniques, history, biological diversity and ethnic diversity will be required for guides and others who will work with tourists. Job training also will be needed for restaurant cooks, service people, drivers, storekeepers and others involved in the hospitality industry. All employees should be regarded as guides/cultural interpreters, regardless of the duties they perform. They must be from, and remain rooted in, the local communities and be intimately familiar with local culture and traditions. They should be made stakeholders, not just wage earners, thus changing the relationship between them and visitors from one of service provider/client to host/guest, allowing for spontaneous interaction and genuine and authentic experiences.

Research

1) **Determine the effects of pesticide/herbicide use in rice fields on amphibian populations.** For example, two separate, 10-to-20-hectare rice fields could be established in Datang and grown for 5-10 years with no agrochemical application. This would (i) provide an exceptional opportunity to determine the effects of these chemicals on amphibians that live and breed in the rice fields; (ii) bring local villagers and farmers of Datang into the process of conserving their natural heritage; (iii) provide important information on the relative importance of pesticides versus frogs for controlling pest insects (rice field frogs eat huge quantities of economically destructive insects); and (iv) provide an economic cost/benefit analysis of agriculture with and without agrochemicals. (Note: local farmers would need to receive financial support for the program.)

2) **Determine the impact of harvesting wild mushrooms and identify strategies to enable local people to use them for food and to generate income.** Once the impact is understood, develop appropriate management and monitoring plans to regulate this activity.

3) **Determine seasonal variation in the elevational distributions and abundance of forest and high-elevation scrub birds.** Although altitudinal migration occurs for a number of species, details are poorly known. Loss of low-elevation habitats may make the knowledge of such details crucial for successful conservation of the avifauna. In other tropical areas of similar latitudes, altitudinal migrations are more pronounced in frugivorous and nectivorous species, which may have implications for seed dispersal or pollination.

4) **Determine habitat-use patterns in mid-montane forest birds.** This may also have implications for management.

5) **Determine the basic ecology of threatened mammals, to develop appropriate conservation strategies.** Most important for such studies are *Macaca arctoides*, *M. mulatta*, *Semnopithecus phayrei*, *Hylobates hoolock*, *Selenarctos thibetanus*, *Ailurus fulgens*, *Panthera pardus*, *Panthera tigris*, *Cervus unicolor*, *Budorcas taxicolor*, *Capricornis sumatraensis*, and *Naemorhedus caudatus*.

6) **Conduct molecular systematic research in reptiles and amphibians to understand the levels of geographic variation and possible cryptic species.**

7) **Conduct detailed taxonomic research on the macrofungi growing in the region.** Many of the reported species listed under a name being used for European or North American fungi are in reality different, oftentimes new, species. This has important conservation implications since some currently unknown number of the reported Gaoligongshan species are likely to be either endemic to the region, or at least to Asia, and thus critical to maintain at healthy population levels. Additionally, the fungi of Gaoligongshan are critical to understanding the biogeography of macrofungi since the area has both tropical and temperate species and also species known only from further east or further west.

8) **Research phenology and fruit production of important tree species; study the recruitment and growth of economic species; study the spread of invasive plants.** These data will contribute to appropriate management plans.

9) **Conduct further research about the cultural and social traditions of local villagers.** More in-depth study of the history, culture and ethnic groups of Baihualing Village and of other villages at the base of the reserve, will provide a base of information for educating local residents and visitors and will provide background for cultural preservation strategies. In addition, records from the Ming Period about the Gaoligonshan area and archaeological investigation along the Southern Silk Road will reveal important information about the region's history and landmarks.

Further inventories

1) **Inventory cultural assets and historial resources in depth.**

2) **Inventory key organisms in the disturbed, lower elevations, to identify high-quality areas for investment in restoration and protection.**

3) **Take quantitative samples to estimate populations of key species.** This will be especially important for species vulnerable to disturbance. An understanding of population sizes will help guide management decisions.

4) **Fill the prominent gaps in geographic and taxonomic coverage of inventories to date.** Focus particularly on the following:

   a. **Herpetological inventories, especially for frogs at high elevations and for snakes in the reserve.** Virtually the only way to sample lizards and snakes effectively is to use aluminum drift-fences with associated traps.

   b. **Inventories of the avifauna above 3,000 m.**

   c. **An expanded program of botanical exploration.** Areas away from Baihualing Station and the east slope along the Southern Silk Road are not known botanically. In particular, inventories should focus on the poorly known, west-facing slopes of the range, where soil conditions and rainfall patterns are different from the east.

   d. **Macrofungal inventories.** Our limited collecting indicates that only a small portion of the macrofungal diversity has been documented. Inventories using both opportunistic and quantitative techniques also will reveal community composition, distribution patterns, and potential specificity of host, site, and substrate.

Monitoring

1) **Measure the effectiveness of ecologically sensitive ecotourism.** Variables to measure include impact on biological and cultural communities, economic gains to the reserve and neighboring communities, and sustainability of the program. Participation of village residents in planning and implementing these monitoring projects will be crucial to success.

2) **Monitor pheasant populations through regular censuses.** Hunting does not seem to be a pressure on these populations at present, but poaching could be a long-term threat.

3) **Monitor populations of forest birds (*Alcippe*, forest understory thrushes, flycatchers, laughing-thrushes) in disturbed buffer areas.**

4) **Monitor amphibians to determine population trends.** If populations are declining within the reserve, conduct research to determine why they are declining and build management plans to abate the causes.

5) **Monitor populations (abundance, size of fruiting bodies, fruiting season) of some of the macrofungi being harvested for food and market.** This will provide useful information to manage the harvesting of wild mushrooms to preserve the populations of economically important species.

6) **Monitor populations of threatened mammals.**

附　录　Appendices

2002年6月17-26日在高黎贡山国家级自然保护区开展的快速生物调查中记录到的大型真菌种类
（最新资料请见 *www.fieldmuseum.org/rbi*）调查组成员：*Greg Mueller* 杨 斌 葛尚义

真菌 FUNGI

属 Genus	种 Species	百花岭 Baihualing	大塘 Datang	赧亢 Nankang
Agaricus	sp. 1	1	–	–
Amanita	*pantherina* group (new sp.)	1	–	–
Amanita	*rubrovolvata*	1	–	–
Amanita	*vaginata* cf.	–	2	–
Amanita	*subglobosa*	1	–	–
Amanita	sp.	1	–	–
Amanita	*sinensis*	–	–	–
Amanita	*fritillaria*	–	2	–
Armillaria	*tabescens*	1	–	–
Armillaria	sp.	–	2	–
Auricularia	*auricula-judea*	1	–	–
Auricularia	*delicata*	1	2	–
Auriscalpium	*vulgare*	–	2	–
Boletellus	*annanus* s.l.	1	–	–
Boletus	*ornatipes*	1	–	–
Boletus	sp. 1	1	–	–
Boletus	sp. 2	1	–	–
Boletus	*sinoaurantiacus*	–	2	–
Boletus	*bicolor*	–	2	–
Boletus	sp. 3	1	–	–
Boletus	sp. 4	1	–	–
Boletus	sp. 5	–	2	–
Boletus	sp. 6	–	2	–
Boletus	sp. 7	–	2	–
Boletus	sp. 8	–	2	–
Boletus	sp. 9	–	2	–
Boletus	sp. 10	–	2	–
Boletus	sp. 11	–	2	–
Boletus	sp. 12	–	2	–
Boletus	sp. 13	–	2	–
Boletus	sp. 14	–	2	–
Campanella	*junghuhnii*	1	–	–
Cantharellus	*cibarius*	–	2	–
Clavicorona	*pyxidata*	1	–	–
Cortinarius	*iodes*	–	2	–
Cortinarius	sp.	–	2	–
Crepidutos	sp.	–	2	–
Cyathus	sp. 1	1	–	–
Cyathus	sp. 2	1	2	–
Cyathus	sp. 3	–	–	3
Cyptotrama	*asperata*	–	2	–
Dictyopanus	sp. 1	1	–	–
Dictyopanus	sp. 2	1	–	–
Discomycete	–	1	–	–
Filoboletus	sp.	1	–	–

Species of macrofungi recorded during a rapid biological inventory of the Gaoligong Mountain National Nature Reserve, Yunnan, China, 17-26 June 2002. Fungi inventory team: Gregory Mueller, Yang Bin, Shangyi Ge. Updated information will be posted at www.fieldmuseum.org/rbi.

真菌 FUNGI				
属 Genus	种 Species	百花岭 Baihualing	大塘 Datang	赧亢 Nankang
Fistulina	hepatica	–	2	–
Galiella	celebica cf.	–	2	–
Geaster	triplex	1	–	–
Gomphidius	viscidus	1	–	–
Hebeloma	sp.	1	–	–
Hydnum	repandum	–	2	–
Hygrophoropsis	aurantiaca	1	–	–
Hygrophorous	speciosus cf.	1	–	–
Inocybe	sp.	1	–	-
Laccaria	amethystina	1	–	–
Laccaria	bicolor	1	–	–
Laccaria	laccata	1	–	3
Laccaria	sp. 1	1	–	3
Laccaria	striatula?	1	2	3
Laccaria	sp. 2	1	–	3
Laccaria	sp. 3	–	–	3
Lactarius	lignyotus	1	–	–
Lactarius	volemus cf.	–	2	–
Lactarius	sp. 1	1	–	–
Lactarius	sp. 2	–	2	–
Lactarius	sp. 3	–	–	3
Lactocollybia	sp.	1	–	–
Laetiporus	sp.	1	–	–
Laetiporus	sulphureus	–	2	–
Leccinum	virens	–	2	–
Leccinum	sp.	1	–	–
Lentinula	boryana	1	2	–
Lepiota	sp. 1	1	–	–
Lepiota	sp. 2	–	2	–
Lepiota	sp. 3	–	2	–
Leucoagaricus	rubrotincta	–	–	3
Marasmiellus	nigripes	1	–	–
Marasmiellus	sp. 1	–	2	–
Marasmiellus	sp. 2	1	–	–
Marasmius	purpureostriatus	1	–	–
Marasmius	sp. 1	1	–	–
Marasmius	sp. 2	1	–	–
Marasmius	sp. 3	1	–	–
Marasmius	sp. 4	1	–	–
Megacollybia	platyphylla	1	–	–
Melanoleuca	sp.	1	–	–
Mycena	haematopus	1	2	–
Mycena	sp. 1	1	2	–
Mycena	sp. 2	1	–	-
Mycena	sp. 3	1	–	–

真菌 FUNGI

属 Genus	种 Species	百花岭 Baihualing	大塘 Datang	赧亢 Nankang
Mycena	sp. 4	1	–	–
Mycena	sp. 5	1	–	–
Mycena/Omphalina	–	1	–	–
Mycena/Pholiota	–	1	–	–
Omphalina	sp.	1	2	–
Oudmansiella	*yunnanensis*	–	2	–
Oudmansiella	*muscida*	1	–	3
Panus	*rudis*	1	2	–
Paxillus	*involutus*	1	–	–
Peziza	sp.	1	–	–
Phaeocollybia	sp.	–	2	–
Phylloporus	*rhodoxanthus*	1	2	–
Phylloporus	sp. 1	–	2	–
Phylloporus	sp. 2	–	2	–
Phylloporus	sp. 3	–	–	3
Pluerotus	sp.	–	–	3
Pluteus	*cervinus* cf.	–	–	3
Pluteus	sp.	1	–	–
Polyporus	*alveolaris*	1	–	–
Polyporus	*varius*	1	–	–
Psathyrella	sp. 1	–	2	–
Psathyrella	sp. 2	–	2	–
Rhodocollybia	sp. 1	1	2	–
Rhodocollybia	sp. 2	1	–	–
Rozites	*emodensis*	–	2	–
Russula	–	–	2	–
Russula	*foetens* group 1	1	2	3
Russula	*foetens* group 2	–	2	3
Russula	*foetens* group 3	1	–	–
Russula	sp. 1	1	–	–
Russula	sp. 2	1	–	–
Russula	sp. 3	1	–	–
Russula	sp. 4	–	2	–
Russula	sp. 5	–	2	–
Russula	sp. 6	–	2	–
Russula	sp. 7	–	–	3
Russula	sp. 8	–	–	3
Sarcodon	sp.	–	2	–
Sarcoscypha ?	sp. 1	–	2	–
Sarcoscypha	sp. 2	–	2	–
Schizophyllum	*commune*	1	–	–
Scleroderma	*areolatum*	1	–	–
Sparassis	*crispa* cf.	–	2	–
Steccherinum	*ochraceum*	1	–	–
Strobilomyces	*floccopus*	1	–	–

真菌 FUNGI				
属 Genus	种 Species	百花岭 Baihualing	大塘 Datang	赧亢 Nankang
Stropharia	sp.	–	2	–
Suillus	sp. 1	1	–	–
Suillus	sp. 2	1	–	–
Suillus	sp. 3	–	2	–
Suillus	sp. 4	–	2	–
Thelephora	sp.	1	–	–
Tremella	*pulvinalis*	–	2	–
Tricholoma	*flavovirens*	1	–	–
Tricholoma	sp.	1	–	–
Tylopilus	*virens*	–	2	–
Tylopilus	sp.	–	2	–
Tyromyces	*chioneus*	1	–	–
Xeromphalina	*temuipes*	1	2	–
Xerula	*furfuracea*	1	2	–
Xylaria	sp.	1	–	–

蕨类 FERNS

科/种 Family/Species
Huperziaceae
Huperzia delavayi (Christ & Herter) Ching
H. serratum (Thunb.) Trev
Lycopodiaceae
Diaphastrum complanatum (L.) Holub
Lycopodium clavatum L.
L. zonatum Ching
Selaginellaceae
Selaginella involvens (Sw.) Spring.
S. leptophylla Bak.
S. monospora Spring.
Equisetaceae
Equisetum arvense L.
Hippochaete hiemale (L.) Borner
Botrychiaceae
Botrypus lanuginosum (Wall.) Holub
B. virginianus (L.) Holub
Sceptridium robustum (Supr.) Ching
Angiopteridaceae
Angiopteris esculenta Ching
Osmundaceae
Osmunda japonica Thunb.
Plagiogyriaceae
Plagiogyria communis Ching
P. glaucescens Ching var. *arguta* Ching
P. media Ching
Gleicheniaceae
Dicranoteris dichotoma (Thunb.) Bernh.
Diplopterygium giganteum (Wall. ex Hook.) Nakai
Lygodiaceae
Lygodium japonicum (Thunb.) Sw.
Hymenophyllaceae
Crepidomanes latealatum (V. d. B) Cop.
Mecodium badium (Hook. & Grev.) Cop.
M. exsertum (Wall. ex Hook.) Cop.
M. levingei (Clarke) Cop.
M. microsorum (V. d. B.) Ching
Dicksoniaceae
Cibotium barometz (L.) J. Sm.
Cyatheaceae
Alsophila spinulosa (Hook.) Tryon

科/种 Family/Species
Dennstaedtiaceae
Dennstaedtia scabra (Wall.) Moore
Microlepia khasiyana (Hook.) Presl
M. marginata (Houtt.) C. Chr.
M. pilosissima Ching
M. platyphylla (D. Don) J. Sm.
Lindsaeaceae
Lindsaea odorata Roxb.
Lindsaea cultrat (Willd.) Sw.
Stenoloma chusanum (L.) Ching
Pteridiaceae
Pteridium revolutum (Bl.) Nakai
Pteridaceae
Pteris aspericaulis Agardh
P. decrescens Christ
P. majestica Ching ex Ching & S. K. Wu
P. nervosa Thunb.
P. venusta Kunze
P. vittata L.
P. wallichiana Agard.
Sinopteridaceae
Aleuritopteris subrufa (Bak.) Ching
Cheilanthes chusana Hook.
Onychium contiguum Hope
O. japonicum (Thunb.) Kze.
O. japonicum (Thunb.) Kze. var. *lucidum* (D. Don) Christ
O. tenuifrons Ching
Pellaea smithii C. Chr.
Adiantaceae
Adiantum caudatum L.
A. edgeworthii Hook.
A. malesianum Ghatak.
A. philippense L.
Hemionitidaceae
Coniogramme affinis Hieron.
C. caudata (Ettingsh.) Ching
C. procera Fee
Gymnopteris vestita Underw.
Antrophyaceae
Antrophyum coriaceum (D. Don.) Hook. & Bak.
A. obovatum Bak.

Species of ferns recorded during a rapid biological inventory of the Gaoligong Mountain National Nature Reserve, Yunnan, China, 17-26 June 2002. Botanical inventory team: Lilan Deng, Jun Wen, Xiaochuan Shi.

蕨类 FERNS

科/种 Family/Species

Vittariaceae

Vittaria doniana Mett. ex Hieron

V. linearifolia Ching

V. modesta Hand. -Mazz.

Athyriaceae

Allantodia bella (Clarke) Ching

A. himalayensis Ching

A. laxifrons (Rosenst.) Ching

A. megaphylla (Bak.) Ching

A. spectabilis (Presl.) Ching

A. viridissima (Christ) Ching

Athyrium bucahwangense Ching

A. chingianum Z. R. Wang & X. C. Zhang

A. clarkei Bedd.

A. drepanopterum (Kunze) A. Br.

A. fangii Ching

A. fimbriatum Moore

A. himalaicum Ching ex Mehra & Bir

A mackinnoni (Hope) C. Chr.

Lunathyrium medogense Ching & S. K. Wu

Thelypteridaceae

Cyclogramma auriculata (J. Sm.) Ching

Dictyocline griffithii Moore

Glaphylopteridopsis erubescens (Hook.) Ching

Macrothelypteris toresslana (Gaud.) Ching

Metathelypteris laxa (Franch. & Sav.) Ching

Parathelypteris beddomei (Bak.) Ching

Pronephrium penangianum (Hook.) Holtt.

Pseudocyclosorus esquirolii (Christ) Ching

P. repens (Hope) Ching

Pseudophegopteris levingei (Clarke) Ching

P. microstegia (Hook.) Ching

P. pyrrhorachis (Kunze) Ching

Aspleniaceae

Asplenium ensiforme Wall. ex Hook. & Grew.

A. laciniatum D. Don

A. tenuifolium D. Don

A. trichomanes L.

A. unilaterale Lam.

A. varians Wall. ex Hook. & Grev.

Ceterachopsis paucivenosa (Ching) Ching

Neottopteris nidus (L.) J. Sm.

科/种 Family/Species

Onocleaceae

Matteuccia struthiopteris (L.) Todaro

Blechnaceae

Blechnidium melanopus (Hook.) Moore

Woodwardia magnifica Ching & P. S. Chin

Peranemaceae

Acrophorus macrocarpus Ching & S. H. Wu

Diacalpe annamensis Tayawa.

Diacalpe aspidioides Bl.

Dryopteridaceae

Cyrtomium hookerianum (Presl) C. Chr.

C. macrophyllum (Makino) Tagawa

Dryopteris acutodentata Ching

D. conjugata Ching

D. lachoongensis (Bedd.) Nayar & Kaur

D. sinofibrillosa Ching

D. sublacera Christ

D. wallichiana (Spreng.) Hyl.

Leptorumohra quadripinnata (Hayata) H. Ito

Lithostegia foeniculacea (Hook.) Ching

Polystichum acanthopyllum (Franch.) Christ

P. acutidens Christ

P. conaense Ching & S. K. Wu

P. deltodon (Bak.) Diels

P. longipaleatum Christ

P. microchlamys (Christ) Matsum.

P. moupinense (Franch.) Bedd.

P. mucronifolium (Bl.) Presl

P. obliquum (D. Don) Moore

P. punctiferum C. Chr.

P. salwinense Ching & H. S. Kung

P. wattii (Bedd.) C. Chr

P. yigongense Ching & S. K. Wu

Aspidiaceae

Ctenitis apiciflora (Wall. Ex Mett) Ching

C. clarkei (Bedd.) Ching

C. dentisora Ching

C. nidus (Clarke) Ching

Egenolfia appendiculata (Willd.) J.Sm

Tectaria coadunata (Hook. & Grev.) C. Chr.

T. polymorpha (Hook.) Cop.

附录2

蕨类 FERNS

科/种 Family/Species
Elaphoglossaceae
Elaphoglossum conforme (Sw.) Schott
Nephrolepidaceae
Nephrolepis auriculata (L.) Trimen
Oleandraceae
Oleandra intermedia Ching
O. wallichii (Hook.) Presl
Davalliaceae
Araiostegia delavayi (Clarke & Bak.) Ching
A. perdurans (Christ) Cop.
Humata assamica (Bedd.) C. Chr.
H. griffithiana (Hook.) C. Chr.
Leucostegia immersa (Wall. ex Hook) Presl.
Polypodiaceae
Arthromeris himalayensis (Hook.) Ching
A. tenuicauda (Hook.) Ching
A. wardii (Crarke) Ching
Colysis pothifolia (D. Don.) Presl.
Lepisorus bicolor Ching
L. contortus (Christ) Ching
L. elegans Ching & W. M. Chu ex W. M. Chu
L. loriformis (Mett) Ching var. *steniste* (Clarke) Ching
L. marcrosphaerus (Bak.) Ching
L. subconfluens Ching
L. sublinearis (Bak.) Ching
Microsorum henryi (Christ) Kuo
Phymatodes chrysotricha (C. Chr.) Ching
P. crenatopinnata (Clarke) Ching
P. ebenipes (Hook.) J. Sm.
P. griffithiana (Hook.) Pichi-Serm.
P. integerrima Ching
P. malacodon (Hook.) Ching
P. oxyloba (Kunze) Ching
P. rhynchophylla (Hook.) J. Sm.
P. stewartii (Bedd.) Ching
P. stracheyi (Ching) Ching
Polypodium subamoenum Clarke
P. wattii (Bedd.) Tagawa
Pyrrosia costata (Presl ex Bedd.) Tagawa & Iwatsuki
P. heteractis (Mett. ex Kuhn.) Ching
P. tibetica Ching var. *angustata* Ching
Tricholepidium angustifolium Ching
T. tibeticum Ching & S. K. Wu

科/种 Family/Species
Drynariaceae
Drynaria quercifolia (L.) J. Sm.
Pseudodrynaria coronans (Wall.) Ching
Loxogrammaceae
Loxogramme lanceolata (Sw.) Presl
Marsileaceae
Marsilea quadrifolia L.
Azollaceae
Azolla imbricata (Roxb.) Nakai

2002年6月17-26日在高黎贡山国家级自然保护区开展的快速生物调查中记录到的两栖与爬行动物种类调查组

成员: *H. Bradley Shaffer* 王天灿 张 宇

两栖爬行类 AMPHIBIANS AND REPTILES

类群 Taxon	Locality	采集地海拔高度 Elevation Collected (m)	垂直分布范围 Elevational Range (m)	多度 Abundance	活动时间 Time Active
CAUDATA					
Salamandridae					
Tylototriton verrucosus	1,3	1515-2030	1000-2000	H	N (D)
ANURA					
Pelobatidae					
Megophrys lateralis	1	1400? (waterfall)	1000-2000	M	N
M. minor	1	1400-2700	1400-2000	H	N (D)
Scutiger gongshanensis	1	3100	2500-4000	M	N
Bufonidae					
Bufo andrewsi	2	1780	800-3500	L	N
B. burmanus	1	2360	1400-2000	L	N
B. melanostictus	1	1400-1750	640-2000	M	N
Hylidae					
Hyla annectans	2	1785-1980	1500-3000	H	N
Ranidae					
Micrixalus liui	1	2360?	–	L	N
Rana arnoldi	2	1785	1500-2500	L	N
R. grahami	1,2	1400-1980	1500-3000	H	N
R. limnocharis	1,2	1530-1785	640-1500	H	N
R. pleuraden	2,3	1785-2050	1500-2000	H	N
R. yunnanensis	1,2,3	1400-2050	1500-2000	M	N
Rhacophoridae					
Chirixalus doriae	1	1530?	–	L	N
Polypedates megacephalus	1,2,3	1515-2050	640-2000	H	N
LACERTILIA					
Gekkonidae					
Gehyra mutilata	1	1515?	–	H	N
Hemidactylus yunnanensis	1	1515	1300-1540	H	N
Agamidae					
Japalura dymondi	1	2000	1800	L	D
Scincidae					
Sphenomorphus indicus	1	1980-2360	640-1500	M	D

Locality

1 = 百花岭 Baihualing
2 = 大塘 Datang
3 = 赧亢 Nankang

Abundance

H = 高 high
M = 中 medium
L = 低（根据野外考察印象）
low (this is based on
impression in the field)

Time Active

N = 夜间活动的 nocturnal
D = 白天活动的 diurnal
(D) = 只有在下雨天才在白
天活动 active in the daytime
when it is raining only

"采集地海拔高度" 指本次调查中发现或采集到该物种的高度.

"垂直分布范围" 为《高黎贡山国家自然保护区》（薛纪如等，1995）的数据.

Elevation collected refers to the elevation at which the species was encountered during this survey.

Elevational range is that given for the region in Xue 1995.

Species of amphibians and reptiles recorded during a rapid biological inventory of the Gaoligong Mountain National Nature Reserve, Yunnan, China, 17-26 June 2002. Herpetological inventory team: H. Bradley Shaffer, Tianchan Wang, Yu Zhang.

两栖爬行类 AMPHIBIANS AND REPTILES					
类群 Taxon	Locality	采集地海拔高度 Elevation Collected (m)	垂直分布范围 Elevational Range (m)	多度 Abundance	活动时间 Time Active
SERPENTES					
Colubridae					
Amphiesma modesta	3	1980	1100-2100	L	D
Elaphe porphyracea	1	2600	1500-2500	L	D
Ophites laoensis	2	1980?	–	L	D
Ptyas mucosus	2	1785?	–	L	D
Rhabdophis nuchalis	1	1530-2360?	–	L	D
R. subminiatus	1	1530?	–	L	D
Zaocys nigromarginatus	2	1700	1500-2000	L	D

Locality

1 = 百花岭 Baihualing
2 = 大塘 Datang
3 = 報亢 Nankang

Abundance

H = 高 high
M = 中 medium
L = 低（根据野外考察印象）
   low (this is based on
   impression in the field)

Time Active

N = 夜间活动的 nocturnal
D = 白天活动的 diurnal
(D) = 只有在下雨天才在白
   天活动 active in the daytime
   when it is raining only

"采集地海拔高度"指本次调查中发现或采集到该物种的高度.

"垂直分布范围"为《高黎贡山国家自然保护区》（薛纪如等，1995）的数据.

Elevation collected refers to the elevation at which the species was encountered during this survey.

Elevational range is that given for the region in Xue 1995.

附录 APPENDIX 4 2002年6月17-26日在高黎贡山国家级自然保护区开展的快速生物调查中记录到的鸟类
（科的分类、名称和系统顺序与*MacKinnon*和*Phillipps* (2000)同）调查组成员：
Dougas F. Stotz 权瑞昌 李正波 *Debra K. Moskovits*

鸟类 BIRDS

种 Species	多度 Abundance	生境 Habitats	百花岭 Baihualing	大塘 Datang	赧亢 Nankang	海拔高度 Elevation (m)
Phasianidae						
Arborophila torqueola	U	F	X	X	X	2100-2700
Bambusicola fytchii	U	F	X	–	–	2300-2400
Picidae						
Picumnus innominatus	R	F	X	–	–	1500
Picoides canicapillus	U	F	X	–	–	1600-2000
Picoides cathpharius	R	F	–	X	–	2100
Picoides darjeelensis	R	F	–	–	–	2750
Celeus brachyurus	R	F	X	–	–	2100
Picus flavinucha	R	F	X	–	–	1700
Blythipicus pyrrhotis	U	F	X	X	–	1500-2100
Megalaimidae						
Megalaima virens	C	F	X	X	X	1500-2100
Megalaima lineata	R	F	X	–	–	1600
Megalaima franklini	C	F	X	X	X	1500-2300
Megalaima asiatica	U	F	X	–	–	1500-1700
Trogonidae						
Harpactes erythrocephalus	U	F	X	–	–	1600-2100
Cuculidae						
Clamator jacobinus	R	Fe	X	–	–	1900
Cuculus sparverioides	C	F	X	X	X	1500-2350
Cuculus saturatus	U	F	X	X	–	1750-2500
Eudynamys scolopacea	R	S	–	X	–	1750
Psittacidae						
Psittacula finschii	U	P	X	–	–	1500
Apodidae						
Collocalia brevirostris	U	O	X	X	–	1500-1950
Cypsiurus balasiensis	R	O	–	X	–	1750
Apus affinis	U	O	–	X	–	1950
Columbidae						
Columba livia	R	V	–	–	X	2000
Columba hodgsoni	U	F	–	X	X	2000-2200
Streptopelia orientalis	U	S	–	X	–	1850
Streptopelia tranquebarica	R	S,P	–	X	–	1850
Treron sphenura	U	F	X	–	–	1600-1800
Accipitridae						
Spilornis cheela	U	F	X	–	–	1950-2000
Accipiter trivirgata	R	F	X	–	–	1600

Abundance
C = 常见 Common
U = 不常见 Uncommon
R = 稀有 Rare

Habitats
F = 森林 Forest
Fe = 林缘 Forest edge
Fs = 水沟溪流边的林子 Forest along streams
H = 高山灌丛 High elevation scrub
O = 天空 Overhead
P = 牧场和农地 Pastures and agricultural lands
R = 水沟和溪流 Streams and rivers
S = 次生灌丛 Secondary scrub
V = 村子里 Villages

附录 4

Species of birds recorded during a rapid biological inventory of the Gaoligong Mountain National Nature Reserve, Yunnan, China, 17-26 June 2002. Bird inventory team: Dougas F. Stotz, Ruichang Quan, Zhengbo Li, Debra Moskovits. Family level taxonomy, nomenclature and systematic order follow MacKinnon and Phillipps (2000).

鸟类 BIRDS

种 Species	多度 Abundance	生境 Habitats	百花岭 Baihualing	大塘 Datang	赧亢 Nankang	海拔高度 Elevation (m)
Accipiter nisus/virgata	R	F	X	–	–	1600
Aquila chrysaetos	U	O	X	X	–	1600-2000
Spizaetus nipalensis	R	F,O	–	X	–	1950
Ardeidae						
Egretta garzetta	U	P	–	X	–	1750
Bubulcus ibis	U	P	–	X	–	1750
Eurylaimidae						
Psarisomus dalhousiae	R	F	X	–	–	1550
Irenidae						
Chloropsis hardwickii	U	F	X	X	–	1600-1850
Corvidae, Corvinae						
Urocissa flavirostris	U	Fe	X	–	–	1600
Urocissa erythrorhynchus	U	F	–	X	–	1900-2100
Dendrocitta formosae	C	Fe	X	X	–	1500-2000
Pica pica	R	P	X	–	–	1600
Corvus macrorhynchus	U	F,S	–	X	–	1750-2000
Artamus fuscus	R	P	X	–	–	2100
Oriolus traillii	R	Fe	–	X	–	1900
Coracina macei	R	F	X	–	–	2500
Coracina melaschistos	R	Fe	–	X	–	1900-2100
Pericrocotus ethologus	R	Fe	–	X	–	2100
Pericrocotus brevirostris	U	F	X	–	–	1500-1700
Pericrocotus flammeus	U	Fe	X	X	–	1500-2100
Hemipus picatus	R	F	X	–	–	2300
Corvidae, Dicrurinae						
Rhipidura hypoxantha	C	F	X	–	X	1700-2100
Rhipidura albicollis	C	F	X	X	X	1500-2100
Rhipidura aureola	U	F	X	–	–	1500-1600
Dicrurus macrocercus	C	S,P,Fe	X	X	–	1500-2000
Dicrurus aeneus	R	Fe	X	–	–	1600
Dicrurus hottentotus	R	Fe	X	–	–	1500-1600
Cinclidae						
Cinclus pallasii	R	R	–	X	–	1900
Muscicapidae, Turdinae						
Monticola rufiventris	U	F	–	–	X	2000-2100
Myiophonus caeruleus	C	Fs,S	–	X	X	1750-2000
Zoothera mollissima	R	H	X	–	–	3100
Zoothera dauma	U	F	–	X	X	1950-2200
Turdus dissimilis	R	F	–	–	X	2000
Turdus pallidus	R	S	–	X	–	1850
Brachypteryx montana	U	F	X	X	–	2000-2350
Muscicapidae, Muscicapinae						
Muscicapa sibirica	U	Fe	–	X	–	2100
Ficedula hodgsonii	U	F	X	–	X	1700-2100
Ficedula hyperythra	R	F	X	–	–	1600-1800

鸟类 BIRDS

种 Species	多度 Abundance	生境 Habitats	百花岭 Baihualing	大塘 Datang	赧亢 Nankang	海拔高度 Elevation (m)
Ficedula westermanni	U	S,Fe	–	X	–	2000-2300
Ficedula sapphira	R	F	–	X	–	2100
Eumyias thalassina	C	Fe,S	X	X	–	1500-2300
Niltava grandis	R	F	X	–	–	1500-1700
Niltava macgrigoriae	C	F	X	–	–	1600-2000
Niltava sundara	U	F	X	–	–	1800
Cyornis unicolor	R	F	–	X	–	2200
Culicicapa ceylonensis	C	F,S	X	X	X	1500-2350
Tarsiger cyanurus	R	F	–	–	X	2000
Tarsiger indicus	R	F	X	–	–	2300
Copsychus saularis	U	S,P	–	X	–	1750
Phoenicurus frontalis	R	H	X	–	–	3100
Rhyacornis fuliginosa	C	Fs	X	X	X	1500-2100
Cinclidium leucurum	C	F	X	–	–	2400
Enicurus scouleri	U	R	–	X	–	1850
Enicurus immaculatus	U	R	X	X	–	1500-1950
Enicurus schistaceus	R	R	–	X	–	1850
Enicurus leschenaultii	R	Fs	X	–	–	1800
Enicurus maculatus	C	F,Fs	X	X	–	1800-2300
Saxicola torquata	C	P	X	–	–	1750-1850
Saxicola ferrea	C	P,S	X	X	X	1500-2200
Sturnidae						
Sturnus nigricollis	U	P	–	X	–	1750
Acridotheres cristatellus	R	P	–	X	–	1750
Sittidae						
Sitta nagaensis	C	F	X	X	X	1500-2300
Sitta himalayensis	R	F	X	–	–	2300
Sitta yunnanensis	R	Fe	–	X	–	2300
Certhiidae						
Certhia discolor	R	F	X	–	–	2300
Paridae, Remizinae						
Cephalopyrus flammiceps	R	F	X	–	–	2350
Paridae, Parinae						
Parus major	C	Fe,S,P,V	X	X	X	1500-2100
Parus monticolus	C	F,S	X	X	X	1600-2100
Parus spilonotus	C	F	X	X	X	1500-2350
Silviparus modestus	U	F	X	X	X	2000-2350

Abundance

C = 常见 Common
U = 不常见 Uncommon
R = 稀有 Rare

Habitats

F = 森林 Forest
Fe = 林缘 Forest edge
Fs = 水沟溪流边的林子
    Forest along streams
H = 高山灌丛 High elevation scrub
O = 天空 Overhead

P = 牧场和农地 Pastures and
    agricultural lands
R = 水沟和溪流 Streams and rivers
S = 次生灌丛 Secondary scrub
V = 村子里 Villages

鸟类 BIRDS

种 Species	多度 Abundance	生境 Habitats	百花岭 Baihualing	大塘 Datang	赧亢 Nankang	海拔高度 Elevation (m)
Aegithalidae						
Aegithalos concinnus	C	F,S	X	X	X	1600-2100
Hirundinidae						
Hirundo rustica	C	O,P,V	–	X	X	1750-2100
Hirundo daurica	C	P,V	–	X	–	1750
Hirundo striolata	C	P,V	–	X	–	1750
Delichon nipalensis	U	O	–	X	–	1950
Pycnonotidae						
Spizoxis canifrons	R	F	X	–	–	1700
Pycnonotus striatus	R	F	X	–	–	2000
Pycnonotus xanthorrhous	C	P,S,V	X	X	–	1500-1750
Pycnonotus aurigaster	C	P,S	X	–	–	1500
Pycnonotus flavescens	U	F	–	–	X	2000
Hypsipetes mcclellandii	C	Fe	X	X	–	1800-2200
Hypsipetes leucocephala	C	F,S	X	X	–	1500-2000
Cisticolidae						
Prinia atrogularis	U	S	X	–	X	1500-2000
Prinia hodgsonii	C	P	X	–	–	1500
Zosteropidae						
Zosterops japonicus	C	P,V	X	X	–	1500-1850
Sylviidae, Acrocephalinae						
Tesia castaneoventer	U	F	X	X	X	2100-2800
Tesia cyaniventer	C	F		X	X	2000-2400
Cettia fortipes	C	S	–	X	–	1750-2300
Phylloscopus armandii	U	F	X	X	–	2350-2700
Phylloscopus maculipennis	U	F	X	X	X	2100-2800
Phylloscopus davisoni	C	F	X	X	X	1500-2700
Seicercus burkii	C	S,Fe	X	X	X	1750-2300
Seicercus castaniceps	C	F	X	X	X	1850-2350
Abroscopus schisticeps	R	F	X	–	–	2300
Tickellia hodgsoni	R	F	X	–	–	2700
Sylviidae, Megalurinae						
Megalura palustris	R	P	X	–	–	1550
Sylviidae, Garrulacinae						
Garrulax albogularis	U	F	–	X	–	2200
Garrulax affinis	U	F	X	–	–	3000
Garrulax erythrocephalus	R	F	–	–	X	2100
Garrulax canorus	C	F,S	X	X	X	1600-2200
Garrulax milnei	U	F	X	X	–	1800-1900
Sylviidae, Sylviinae						
Pomatorhinus ferruginea	R	F	X	–	–	1800
Napothera epilepidota	R	Fs	–	–	X	2100
Sphenocichla humei	R	F	–	X	–	2600
Stachyris rufifrons	R	F	–	–	X	2100
Stachyris ruficeps	U	F	X	–	X	1500-2000

鸟类 **BIRDS**

种 Species	多度 Abundance	生境 Habitats	百花岭 Baihualing	大塘 Datang	赧亢 Nankang	海拔高度 Elevation (m)
Stachyris chrysaeus	U	F	X	–	–	1500-1600
Babax lanceolatus	R	S	–	X	–	1850
Leiothrix argentauris	U	F,S	X	–	–	1500-1600
Leiothrix lutea	U	S,F	X	X	–	1800-2000
Cutia nipalensis	U	F	X	X	–	2300-2350
Pteruthius rufiventer	R	F	X	–	–	2350
Pteruthius flaviscapis	U	F	X	X	X	1600-2100
Pteruthius melanotis	R	F	X	X	–	1600-2350
Actinodera egertoni	C	Fe	X	–	–	1500-2000
Minla cyanuroptera	C	F,S	X	X	X	1500-2350
Minla ignotincta	C	F	X	X	–	1700-3100
Minla strigula	C	F	X	X	X	1950-3000
Alcippe cinerea	U	F	–	X	X	2000-2300
Alcippe castaneceps	C	F	X	X	X	1850-2300
Alcippe vinipectus	U	F	X	X	–	1850-2300
Alcippe ruficapilla	R	F	–	–	X	2100
Alcippe dubia	U	F	X	X	X	2100-2700
Alcippe morrisonia	C	F	X	X	X	1500-2350
Myzornis pyrrhoura	U	H	X	–	–	3100
Heterophasia melanoleuca	C	F	X	X	X	1500-2350
Heterophasia pulchella	C	F	X	X	X	1500-2100
Yuhina castaneceps	C	S,F	X	–	–	1500-1700
Yuhina flavicollis	C	F,S	X	X	X	1500-2350
Yuhina gularis	C	H,F	X	X	–	2100-3100
Yuhina diademata	R	F	–	X	–	2500
Yuhina occipitalis	C	F	X	X	–	2300-3100
Yuhina nigrimenta	R	F	X	–	–	1700
Yuhina zantholeuca	U	F,S	X	X	–	1500-1950
Paradoxornis unicolor	U	F	X	–	–	2800
Paradoxornis gularis	U	S	X	–	–	1500-1600
Paradoxornis brunneus	U	S	–	X	–	1850-2100
Paradoxornis nipalensis	R	F	–	–	X	2000
Nectariniidae						
Dicaeum ignipectus	R	Fe	–	X	–	2000
Aethopyga gouldiae	C	Fe,S	X	X	–	2000-3100
Aethopyga nipalensis	C	Fe	X	X	X	1900-2500
Aethopyga saturata	U	Fe,S	X	–	–	1500-1600
Arachnothera magna	R	Fe	X	–	–	1600

Abundance

C = 常见 Common
U = 不常见 Uncommon
R = 稀有 Rare

Habitats

F = 森林 Forest
Fe = 林缘 Forest edge
Fs = 水沟溪流边的林子
　　 Forest along streams
H = 高山灌丛 High elevation scrub
O = 天空 Overhead

P = 牧场和农地 Pastures and
　　agricultural lands
R = 水沟和溪流 Streams and rivers
S = 次生灌丛 Secondary scrub
V = 村子里 Villages

鸟类 BIRDS

种 Species	多度 Abundance	生境 Habitats	百花岭 Baihualing	大塘 Datang	赧亢 Nankang	海拔高度 Elevation (m)
Passeridae, Passerinae						
Passer montanus	C	P	–	X	–	1750
Passer rutilans	C	P	–	X	–	1750-2000
Passeridae, Motacillinae						
Motacilla alba	C	V,P	X	X	X	1500-2100
Anthus roseatus	R	H	X	–	–	3100
Passeridae, Estrildinae						
Lonchura punctulata	R	P	–	X	–	1750
Fringillidae						
Carduelis ambigua	C	S,P,V	–	X	X	1850-2300
Carduelis thibetana	R	H	X	–	–	3100
Pyrrhula erythaca	R	F	–	–	X	2100

Abundance

C = 常见 Common
U = 不常见 Uncommon
R = 稀有 Rare

Habitats

F = 森林 Forest
Fe = 林缘 Forest edge
Fs = 水沟溪流边的林子
     Forest along streams
H = 高山灌丛 High elevation scrub
O = 天空 Overhead

P = 牧场和农地 Pastures and
    agricultural lands
R = 水沟和溪流 Streams and rivers
S = 次生灌丛 Secondary scrub
V = 村子里 Villages

2002年6月17-26日在高黎贡山国家级自然保护区开展的快速生物调查中记录到的大型兽类调查组成员：
艾怀森 蔺汝涛

兽类生境 LARGE MAMMALS

种 Species	生境 Habitats	垂直分布范围 Elevational Range (m)
Insectivora/Erinaceidae		
Erinaceus sp.	Secondary forests	1600-2200
Talpidae		
Uropsilus gracilia	Farmlands, forests	1400-2000
Scandentia/Tupaiidae		
Tupaia belangeri	Trees	1200-2500
Primates/Cercopithecidae		
Macaca mulatta	EBF	1000-2600
Macaca nemestrina	EBF	1800-2400
Macaca assamensis	EBF	1800-2700
Macaca arctoides	EBF	2000-2700
Semnopithecus phayrei	MMHEBF	2000-2800
Hylobatidae		
Hylobates hoolock	MMHEBF	2100-2500
Carnivora/Ursidae		
Selenarctos thibetanus	MMHEBF	1400-2800
Helarctos malayanus	MEBF	1300-1900
Procyonidae		
Ailurus fulgens	MMHEBF	2000-3700
Mustelidae		
Mustela kathiah	Farmlands, village sites	1400-1900
Mustela sibirica	Forests, farming lands, village areas	1000-2400
Melogale moschata	Forest edge and farming lands on the eastern side	1100-1800
Canidae		
Vulpes vulpes	EBF	1700-2500
Cuon alpinus	EBF	1700-2500
Felidae		
Prionailurus bengalensis	All over the reserve	800-2800
Catopuma temmincki	EBF	1500-3000
Panthera pardus	MMHEBF	2200-3000

Habitats

All over the reserve = 整个保护区
and above = 及其以上地区
 (referring to elevation)
Bamboo forests = 竹林
Broadleafed forests = 阔叶林
EBF = 常绿阔叶林 Evergreen broadleafed forests
Farmlands = 农地
Forest Edge = 林缘
Forests = 森林

MMHEBF = 中山湿性常绿 阔叶林 Middle-mountain humid evergreen broadleafed forests
MEBF = 季风常绿阔叶林 Monsoon evergreen broadleafed forests
on the eastern side = 东坡
Secondary forests = 次生林
Trees = 乔木
Villages = 村子里

Species of large mammals recorded during a rapid biological inventory of the Gaoligong Mountain National Nature Reserve, Yunnan, China, 17-26 June 2002. Mammal inventory team: Huaisen Ai, Rutao Lin.

兽类生境 LARGE MAMMALS

种 Species	生境 Habitats	垂直分布范围 Elevational Range (m)
Viverridae		
Viverra zibetha	MMHEBF	1800-2500
Prionodon pardicolor	MMHEBF	1800-2500
Paguma larvata	EBF	1500-2600
Artiodactyla/Bovidae		
Capricornis sumatraensis	MMHEBF	2000-2800
Naemorhedus caudatus	MMHEBF	2000-2800
Budorcas taxicolor	MMHEBF and above	Above 2200
Cervidae		
Muntiacus vaginalis	All over the reserve	800-2800
Cervus unicolor	MMHEBF	2000-2800
Moschidae		
Moschus berezovskii	MMHEBF	2000-2800
Suidae		
Sus scrofa	MMHEBF	1900-2500
Pholidota/Manidae		
Manis pentadactyla	EBF	1000-2000
Lagomorpha/Leporidae		
Lepus comus	All over the reserve	800-2600
Rodentia/Sciuridae		
Ratufa bicolor	EBF	1300-2000
Callosciurus erythraeus	All over the reserve	1500-2500
Tamiops swinhoei	EBF	1000-1800
Dremomys pemyi	EBF	1300-2000
Hylopetes alboniger	Broadleaf forests	1900-2800
Muridae		
Rhizomys pruinosus	Bamboo forests	1400
Hystricidae		
Hystrix hodgsoni	Forests, farming lands	1700-2400
Hystrix yunnanensis	Forests, farming lands	1400-2000
Atherurus maevourus	Forests, farming lands	1500-2000

Habitats

All over the reserve = 整个保护区

and above = 及其以上地区
(referring to elevation)

Bamboo forests = 竹林

Broadleafed forests = 阔叶林

EBF = 常绿阔叶林 Evergreen broadleafed forests

Farmlands = 农地

Forest Edge = 林缘

Forests = 森林

MMHEBF = 中山湿性常绿阔叶林 Middle-mountain humid evergreen broadleafed forests

MEBF = 季风常绿阔叶林 Monsoon evergreen broadleafed forests

on the eastern side = 东坡

Secondary forests = 次生林

Trees = 乔木

Villages = 村子里

社会文化资源

本附录各表格总结了考察人员（Anne Underhill 和 Victoria Drake）于2002年7月在百花岭村开展的社会和文化资源快速调查中所收集到的情况。调查对象包括百花岭行政村村长（村民委会主任）、各自然村（村民小组）组长以及每个自然村中随即挑选的一户家庭

调查中使用的问题

01 你家有几口人？在此地定居的时间有多久了？有几个孩子？

02 你家是什么民族？

03 目前在此地仍然还在庆祝的地方节日有那些？有那些信仰等？

04 你家耕种多少土地？水田和旱地分别几亩？种何种作物？

05 家庭年均收入是多少？每一种作物的收入分别是多少？

06 家里使用的主要燃料/灶是什么？怎样采集薪材？

07 家里还有其它何种赚钱的方式？

08 家里有没有人在政府部门工作？

09 最大的家庭开支是什么？在村里你最关心的问题是什么？

10 你听没听说过农民示范项目？你觉得该项目怎么样？

11 你家家庭成员受教育的程度如何？你对正在修建的学校有何看法？你一年的教育附加费是多少？（大部分人回答：一年50元）

12 村里的人口增长情况如何？

13 对发展旅游（其影响以及参与的方式）有何意见和看法？

14 当地尚有何种手工艺？

15 你认为发展整个村子，最优先考虑的应该是什么？（先进的农业技术、教育和医疗保健是最该优先考虑项目）

16 你是否认为当地的环境保护与庄稼的收成和薪材的资源有直接的联系？（大部分人对该问题的回答是肯定的

1 - 汉龙村
2 - 大鱼塘上社
3 - 大鱼塘下社
4a - 邦迈
4b - 古兴寨
5 - 桃园村
6a - 老蒙寨
6b - 百花岭
6c - 麻栎山
7 - 芒岗
8 - 芒晃

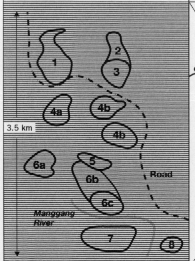

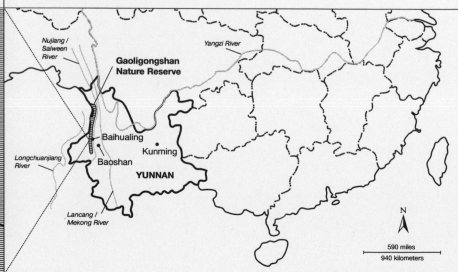

表1　百花岭各自然村社会文化总结（基本按从北到南，从高海拔到底海拔的顺序）

社会文化资源				
村名	户数	人口	土地面积	民族
汉龙村	40多户	250	200多亩水田 400多亩旱地	汉族80% 白族10% 傈僳族10%
大鱼塘上社 大鱼塘下社	数据不详 41户	150-200人 178人	数据不详 数据不详	两个村子都是汉族占75%， 白族和傈僳族占25%
邦迈-古兴寨	168户	数据不详	200亩水田 80亩旱地	汉族65%； 白族和傈僳族35%
桃园村	37	150人	120亩水田 80亩旱地	汉族80% 白族10% 傈僳族5% 彝族5%
老蒙寨-百花岭	96	400多人	老蒙寨：326亩； 百花岭：水田和旱地 各 400亩；	老蒙寨大部分是汉族， 也有白族；百花岭大部分 是汉族，有一些白族和极 少数傈僳族、彝族和傣族
麻栎山	78		麻栎山: 500多亩旱地	麻栎山大部分是汉族，也 有傈僳族、白族和彝族
芒岗	108	数据不详	数据不详	汉族60% 傈僳族30% 彝族10%
芒晃	37	140人	数据不详，但旱地明 显比水田多	大部分是汉族，有部分白 族和彝族

注：各村的民族组成估计数字差异很大，本表数据主要由对村子里情况比较熟悉的村民提供。详细情况尚需进一步调查

表2   百花岭各村的经济条件比较：自然资源和面临的挑战

社会文化资源		
村名	自然资源	面临的挑战
汉龙村	附近山上（共3,000多亩）有各种植物资源和可作薪材的灌木林，山上景色优美；同时还有澡堂河温泉；旧街子附近有珍稀的树蕨.	可种植水稻以及甘蔗、咖啡等经济作物的水田较少
大鱼塘上社和下寨	地势较平，交通方便	缺乏清洁的饮用水；以前的鱼塘已经废弃
邦迈-古兴寨	没有什么特别的自然资源	无特别挑战
桃园村	没有什么特别的自然资源	自然条件较差，交通不便，道路陡峭
老蒙寨-百花岭-麻栎山	海拔适中，位于该地区的中心地带	缺乏农业灌溉用水；要走很远的路采集薪材
芒岗	海拔较低，交通较为方便；有两棵大菩提树	离山地自然资源较远；病虫害、鼠患严重
芒晃	海拔较低，交通较为方便	离山地自然资源较远

注：表中关于大鱼塘上寨和下寨的情况主要由下寨的村民提供

社会文化资源	
村子	新的经济活动
汉龙村	场だ家庭种植并出售中草药、板栗，采集蘑菇；利用马匹运送游客和货物；向游客出售食物；一位村民出售羊毛线
大鱼塘上社和下社	有一户人家开了一家小店；有的村民出售中草药、水果、干果和养猪
帮迈-古兴寨	部分家庭制作牛肉干
桃园村	无特别活动
老蒙寨-百花岭-麻栎山	部分农户种植桔子、荔枝、龙眼等经济作物；有的村民在岗党的集市出售猪饲料和猪；而有的则出售鸡；麻栎山有两家小店，出售手工艺品和食品
芒岗	种植果树、核桃和中草药等经济作物
芒晃	有一农户种植葡萄，养兔子,主要供自家食用；有的农户种植芒果、荔枝、龙眼等经济作物，也有养猪的农户；村里有一家小卖部

注：表中大鱼塘上社和下社的情况主要由下社村民提供

社会文化资源		
村子	文化资源	历史资源
汉龙村	附近有观音寺；部分村民会做竹子制品、豆腐和米酒；离规划中的入口区较近	观音寺始建于明代或更早时期；附近有吴家的抗日战争纪念馆；有南方丝绸之路的遗迹；旧街子明代建筑遗址；金厂河附近还有一座废弃的金矿，周围景色秀丽，旁边有几座古墓
大鱼塘上社和下社	有些年纪较大的人（主要是妇女，也有至少一个男子）会做绣花鞋、鞋底，编制竹篮，做葫芦瓢；有的人还会做木雕；村里已经开始建盖一座旅社，计划用来开商店	附近有几处历史遗迹，最早可以追溯到明代；山坡上有抗日战争留下的战壕；一个1932年以前很繁荣的集市；一处抗战时期被日本人破坏的文化宫遗址；当时村里的接龙井也遭破坏；不少人说真正的南方丝绸之路穿过该村的地界
帮迈-古兴寨	有人会酿酒、做绣花鞋、牛肉干、豆制品和织毛衣	缺乏资料
桃园村	少数人会做绣花鞋、荒邓垫 和豆腐；有一村民会弹三弦	缺乏资料
老蒙寨-百花岭-麻栎山	麻栎山有一村民会做披肩和藤椅；百花岭有一妇女会做钱夹；村里有几个石匠，会建盖土木结构的房子；有两人会吹锁呐，其中一人会做；村里还有人会编制竹篮、藤椅和做鞋；有人会做凉鞋；麻栎山有一道很有名的玉米食品；该村还有一人懂传统医术	缺乏资料
芒岗	有人会做酸腌菜；有人会编织；村内有一座傈僳族基督教堂。	缺乏资料
芒晃	有人会制鞋、编制篮子；有位村民会做八宝饭；有人会做卤腐。	缺乏资料

社会文化资源	
村名	组织、机构
汉龙村	共有两户示范家庭和一户试验家庭
大鱼塘上社和下社	原先发起旅游业的几位年长者开始了修建招待所；共有2户示范家庭
帮迈-古兴寨	一户示范家庭
桃园村	一户示范家庭
老蒙寨-百花岭-麻栎山	麦克阿瑟基金会资助项目总部设在该村，建有授课教室一间； 王先生创办的农民生物多样性保护协会在百花岭一带带动了一系列的生态保护项目；百花岭有一户示范家庭
芒岗	共有三户示范家庭和一户试验家庭
芒晃	有位妇女是村里妇联负责人，经常号召人们支持一些公益事业

注：示范家庭和试验家庭项目始于1995年，是农民生物多样性保护协会的一个重要组成部分，受麦克阿瑟基金会资助。
选择示范家庭的标准是看该家庭是否有能力进行投资，开展多种经营活动。他们的开发活动可以得到相关的资金支持。
试验家庭可以从示范家庭那里获得技术支持，但一般没有现金资助。他们向示范家庭学习一些实用的技术和经验

村名	村民对教育的看法
汉龙村	教育很重要，但很少家庭能为孩子支付上高中或上大学的费用
大鱼塘上社和下社	有些村民认为一次性集资建校增加了他们的经济负担；有的则表示；即使孩子上了学，将来的结局也是回家务农，许多家庭表示孩子是家里主要的劳力之一；教师的素质太低，因为大部分教师不愿留在当地任教；本村上中学的孩子比其它村多
帮迈-古兴寨	普遍对学校不能快点建好感到失望，一次性集资也给各个家庭增加了经济负担；村民们普遍认为女孩和男孩的教育一样重要（该村有一女孩正在读大学）；儿子是家里重要的劳动力；教育支出太高，使不少家庭陷入债务；对许多家庭来说，小学教育也是一笔不小的开支，有时一家人辛辛苦苦就只能供一个孩子上学；教学质量差；年轻人需要帮助，了解外地的工作机会；村里部分妇女（各种年龄层次都有）不识字.
桃园村	有些家庭甚至有的孩子，认为集资办学负担太重；有的学龄儿童不识字.
老蒙寨-百花岭-麻栎山	小学急需资金建盖教工和学生宿舍；只有极少数的年轻人上中学；村里没有一个大学生；学费是各个家庭的一大负担；有一村民反映，村里边15%的家庭无力支付孩子上小学的学费；有的孩子生病，得不到及时的医治，影响了他们上学；教师质量亟待提高.
芒岗	村民对此没有表示任何看法.
芒晃	学校离家太远；只有极少数人上中学并完成学业；村里没有大学生；有些妇女不识字

村子	对开发旅游业、自然保护和发展经济的看法
汉龙村	希望在附近的保护区管理站和以后的入口处旅舍与游客有更多的接触；许多家庭表示愿意提供更多旅游服务，但是没有启动资金；可以开发竹类产品，但不宜开发木材产品，因为树木生长期太长；希望能够更多地利用附近的森林资源；大家普遍意识到自然保护的重要性。
大鱼塘上社和下社	村民对游客持欢迎态度，但游客主要集中在汉龙村；应该对希望将来从事旅游活动的人员进行相关培训，例如英语和日语翻译等；开展旅游接待的农户希望向游客出售兰花，但这些需要投资；村民们有一些提供旅游服务的观念；可以选择一些合适的木材开展木雕工艺品制作；如果鱼塘得到修缮的话，村民也可以把鱼卖给游客。
帮迈-古兴寨	有些村民认为旅游与他们生活没有什么联系；无论做什么新的项目，首先都得解决资金问题；各种开支已经把村民们压得透不过气来；村民们可以向游客出售食品和手工艺品；村民们一年到头忙忙碌碌，没有时间和资金进行创新；村民们也可以把马租给游客骑或卖豆制品给他们。
桃园村	村民们对旅游业持欢迎态度，但急需资金进行投资；可以种植经济果木，例如核桃；村民们大部分资金都投到了农业生产上；由于旅游业对村民来说是一件新事业，因此应该进行相关培训。
老蒙寨 -百花岭-麻栎山	需要在成人和儿童中开展环境保护教育；旅游业给村民们带来了接受教育的机会；有的家庭担心旅游业带来的负面影响；有的家庭觉得应该建盖更多的旅馆来接待游客；各个家庭可以向游客出售食品；有的人担心旅游业的大部分收入将归政府所有，而不是反馈到村民中间；有的村民担心学校的建设会被延期；有人则认为别的村子比他们村更有吸引力，游客只会到这些村子去；应该在汉龙村建立一个市场，这样一来，各村村民都可以到那里向游客出售食品；种植果树也可以吸引游客；麻栎山村一位村民反映该村新式的节能灶太少；有的村民则认为市场应该建在百花岭。
芒岗	有的村民对发展旅游业可能带来的收益持怀疑态度，因为他们认为游客多数只会去管理站附近；村民们可以出售野生食用菌和其它食品；有的则认为旅游业会给所有的人带来实惠。
芒晃	有的村民对发展旅游业可能带来的收益持怀疑态度；希望游客多光顾当地的商店；急需资金进行投资；应该在汉龙村一个市场。

SOCIAL AND CULTURAL ASSETS

The following tables summarize the results of the rapid cultural/social asset inventory conducted in Baihauling Village in July 2002. Anne Underhill and Victoria Drake interviewed the village leader, as well as the leader of each hamlet and another family from each hamlet. The list of questions follows.

Questions

Baihualing Village Hamlets

1 - Hanlong
2 - Upper Dayutang
3 - Lower Dayutang
4a - Bangwai
4b - Guxingzhai
5 - Taoyuan
6a - Laomengzhai
6b - Baihualing
6c - Malishan
7 - Manggang
8 - Manghuang

Hamlet, name, ethnic minority

01 How many people live here in your house? How long has your family been in the area? How many children?

02 What is your family's ethnic background?

03 What local festivals, beliefs, spirituality, etc. are still practiced regularly?

04 How much land do you farm? Wet and dry land? What crops in how many mu?

05 What is your average income? How much from which crop(s)?

06 What kind of fuel/stove do you use? How do you collect fuel?

07 What other ways do you make money?

08 Does any family member work for the government?

09 What is your highest cost(s)? concerns for the community?

10 Do you know about the Farmers Demonstration Project? What do you think of it?

11 What is your/family education level? What do you think of the new school being built. How much tax did you pay? (50 yuan per person is standard reply)

12 What is the hamlet population growth?

13 Any opinions or views on tourism, its encroachment, impact, ways to be involved?

14 What local crafts are still in production or can be produced if necessary?

15 What would be your highest priority for improving the hamlet as a whole? (better farming skills-techniques, education & healthcare were top three)

16 Do you think there is a connection between local environmental conservation and the well-being of your crops/fuel stores? (overwhelmingly affirmative)

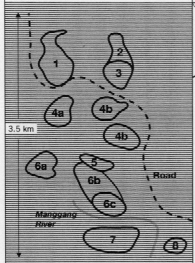

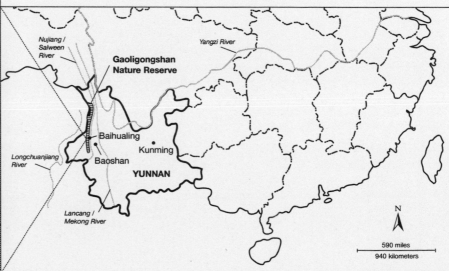

Table 1. Summary of Hamlets in Baihualing Village, from Approximately North to South, and Highest in Elevation to Lowest.

SOCIAL AND CULTURAL ASSETS

		# Families	Population	Total land	Ethnicity
Hanlong		Over 40	c. 250	Over 200 mu* wetland, over 400 mu dry	c. 80% Han c. 10% Bai c. 10% Lisu
Dayutang	*Upper*	Unclear	150-200	Unknown	Both: c. 75% Han 25% Bai and Lisu
	Lower	c. 41	c. 178	Unknown	
Bangwai-Guxingzhai		c. 168	Unknown	c. 200 mu wetland, 80 dry	c. 65% Han, c. 35% Bai and Hui
Taoyuan		c. 37	c. 150	120 mu wetland, 80 dry	c. 80% Han c. 10% Bai 5% Lisu, 5% Yi
Laomengzhai-Baihualing		More than 96	More than 400	400 mu wet, 400 dry	Laomengzhai: mostly Han, also Bai Baihualing: mostly Han, some Bai, a few Lisu, Yi, Dai
Malishan		More than 78	c. 326	over 400 mu wet, over 500 dry	Malishan: mostly Han, Lisu, Bai, Yi
Manggang		c. 108	Unknown	Unknown	c. 60% Han, 30% Lisu, 10% Yi
Manghuang		c. 37	c. 140	Unknown, but more dry than wet land	Mostly Han, some Bai, Yi

* A mu is a Chinese acre, equal to 0.0667 hectares.

Note: Estimates for ethnic groups in each hamlet varied widely. We tried to get figures from knowledgeable people within each hamlet. A more thorough study of the ethnic composition and the cultural practices in the area should be conducted.

Table 2. Economic Differences Between Hamlets in Baihualing Village: Natural Assets and Challenges

SOCIAL AND CULTURAL ASSETS		
	Natural Assets	**Challenges**
Hanlong	Mountain areas nearby (over 3,000 mu) have resources such as wild plants and brush for fuel, and beauty: Zaotanghe hotsprings and waterfalls; rare tree ferns at Jiujiezi	Less low-lying wetland available for rice, sugar-cane and coffee cash crops
Dayutang: Upper, Lower**	Flat setting, easy access to road	Area needs more good-quality water; former fish pond no longer usable
Bangwai-Guxingzhai	Nothing unusual	Nothing unusual
Taoyuan	Nothing unusual	Natural setting makes transportation difficult; very steep road
Laomengzhai-Baihualing-Malishan	Central location at moderate elevation	Area needs more irrigation water for farming; fuel resources far away
Manggang	Lower elevation easier access; 2 beautiful banyan trees here	Distance to mountain resources; serious pest problem: rats, mice
Manghuang	Lower elevation easier access	Distance to mountain resources

***All of our information comes from individuals who live in Lower Dayutang. They tried to generalize for the hamlet as a whole.*

Table 3. Economic Creativity by Hamlet [people taking initiative, creativity, diverse ways to make money; what they are doing already].

SOCIAL AND CULTURAL ASSETS	
	Innovative Economic Activities
Hanlong	Some families grow and sell herbs, nuts; collect wild mushrooms; use their horses to transport tourists and goods, sell foods to tourists; one man sells yarn from sheep.
Dayutang: Upper, Lower	One family has a shop, some sell herbs, fruits, nuts, pigs.
Bangwai-Guxingzhai	Some make dried beef.
Taoyuan	Nothing unusual.
Laomengzhai-Baihualing-Malishan	Some people grow oranges, lychee, longyan; others sell pig feed and pigs in Gandan; some sell chickens; Malishan has 2 stores with craft goods as well as food.
Manggang	Also grow fruits, nuts such as walnuts, herbs.
Manghuang	One family grows grapes, has rabbits for personal consumption; some have mango, lychee, longyan; sell pigs; 1 store.

Table 4. Modern Cultural and Historical Assets by Hamlet.

SOCIAL AND CULTURAL ASSETS

	Modern Cultural Assets	Historical Assets
Hanlong	Near the Guangyin temple; some people know how to make bamboo items; dofu, grain alcohol; close to the planned Gateway Lodge and Visitor Center.	Evidence temple first used in Ming period, or earlier; Mr. Wu's War Museum [Anti-Japanese War, late 1930's-early 40's]; near an actual part of the Southern Silk Road at Jiujiezi with partially standing Ming wall; former location of a gold mine in a beautiful area, at Jingchanghe, with old tombs.
Dayutang: Upper, Lower	Some old people [women and at least one man] can make embroidered shoes and soles, baskets, gourd scoops; some remember wood carving, construction for a guesthouse has begun, planning for shops.	A few historical spots nearby, earliest seem to be Ming: Anti-Japanese War defensive trenches on mountain side; remnants of former thriving town before 1932: a Wenhua Gong Cultural Palace that was destroyed by the Japanese, a Jie Long Qing Welcome Dragon and a Dry Dragon Well also destroyed; some claim that the Southern Silk Road also intersected this area.
Bangwai-Guxingzhai	Some can make alcohol, embroidered shoes, a bean dish, dried beef, sweaters.	Unknown
Taoyuan	A few can still make embroidered shoes, dofu, pickled vegetables; one person can play the sanxian stringed instrument.	Unknown
Laomengzhai-Baihualing-Malishan	Apparently at Malishan at least one person can make horsehair rain capes, wicker stools; one woman at Baihualing made a nice purse using a machine elsewhere; some old men can do stone work for houses and build wooden houses; two can play the suo na trumpet-like instrument, one can make it; others in hamlet can make baskets, wicker stools, shoes; someone can make horsehair saddles; at Malishan there is a traditional Lisu cornmeal dish; one person at Malishan knows traditional medicine.	Unknown
Manggang	Some can make pickled vegetables condiment; some knew how to do weaving with hemp; the Lisu Nationality Protestant Church is here.	Unknown
Manghuang	Some remember how to make shoes, baskets; one can make sticky rice festival dish [baobao fan], some can make pickled dofu.	Unknown

Table 5. Organizational Assets by Hamlet.

SOCIAL AND CULTURAL ASSETS	
	Organizational Asset
Hanlong	2 demonstration families, 1 pioneer family
Dayutang: Upper, Lower	Older men who started the tourism association, began the Guesthouse project; 2 demonstration families.
Bangwai-Guxingzhai	1 demonstration family
Taoyuan	1 demonstration family
Laomengzhai-Baihualing-Malishan	MacArthur Foundation Project Headquarters and classroom; Mr. Wang's office as head of Baihualing village instigated many ecological projects in area; one demonstration family in Baihualing hamlet.
Manggang	3 demonstration families, 2 pioneer families.
Manghuang	One woman was chairperson of a hamlet women's association, encouraging people to donate to good causes.

Note: The demonstration and pioneer families are part of the Gaoligongshan Farmers Biodiversity and Conservation Association [GFBCA] program funded by the MacArthur Foundation and initiated in 1995. Demonstration families may be chosen on the basis of an ability to invest money in diverse farming methods. They are given financial compensation for their efforts. Pioneer families receive technical assistance from demonstration families but no financial assistance. They act as apprentices to the demonstration families.

Table 6. Opinions About Education by Hamlet.

SOCIAL AND CULTURAL ASSETS	
	Opinions About Education
Hanlong	Important, but few can afford tuition for senior high school or university.
Dayutang: Upper, Lower	Some residents expressed a sense of burden because of the one-time tax for school construction. Even if a child has education it is hard to find any job besides farming and families need labor of children for farming; quality of teachers is a big problem—few want to come to area; more young people from Dayutang attend high school here than from other hamlets.
Bangwai-Guxingzhai	Frustration that the school construction is not completed sooner especially because of the one-time tax. Education is important for girls as well as boys [one girl attended college]; the labor of sons especially is needed for farming; tuition fees cause much debt; even the fees for the primary school cause great strain; entire families must make sacrifices for one child to get a good education; teacher quality is a big problem; young people need help in learning about job opportunities outside Baihualing village; several women [all ages] and girls are illiterate.
Taoyuan	Some families, even some with children, feel one-time school construction tax as a burden. Some children are illiterate.
Laomengzhai-Baihualing-Malishan	More money is needed to build houses for teachers and dorms for students for the primary school; only a very few young people go to senior high school and no one has gone to college; any tuition is difficult; one person said 15% of families here cannot afford to send children to primary school; some children have serious health problems that cannot be cured adequately and prevent them from attending primary school; quality of teachers for the local school is a big problem.
Manggang	No opinions expressed.
Manghuang	Distance from the primary school is a big disadvantage for families; few can attend or finish senior high school, no one has gone to college; some women are illiterate.

Table 7. Views About Tourism, Conservation, and Economic Development.

SOCIAL AND CULTURAL ASSETS

	Tourism, Conservation, and Economic Development
Hanlong	Welcome more interaction with tourists at nearby Ranger Station and future Gateway Lodge; families are eager to provide more services for tourists but say they need more start up funds to invest; bamboo products could be used but the trees take too long to grow and should not be used too early; hope to use more nearby forest resources but understand importance of conservation.
Dayutang: Upper, Lower	Tourists are welcome but they all go only to Hanlong; there should be training classes for tourism jobs such as translating English and Japanese; in the future families could sell orchids to visitors, but need money to invest; people have ideas about providing entertainment; people could revive wood carving using suitable wood; they could sell fish from the pond if the pond could be revitalized.
Bangwai-Guxingzhai	Some people feel no connection at all to tourism and cannot imagine how it is relevant to their lives; for any new venture people need more cash to invest; people already have too many basic expenses that are a strain; people could make and sell food items and revive crafts; people have little time or money to consider methods of economic innovation; people could rent horses or sell a bean dish to tourists.
Taoyuan	Tourism is welcome but people need cash to invest; could plant walnut trees; however people need more cash just to put into farming alone; there should be a training class for tourism since many people have no idea what to expect.
Laomengzhai-Baihualing-Malishan	Adults and children need seminars on protection of the environment; tourism provides an opportunity for educational exchange; however some families cannot imagine what impact tourism will have; others say there should be more guesthouses for tourists; families could sell foods to tourists; fear that the government instead of local people will receive profits from tourism; anxiety about delay in school construction; feeling that other hamlets have a better location to attract tourists; there should be a market in Hanlong where people from all hamlets can sell foods to tourists; planting fruit trees would attract tourists too; one person said that few people in Malishan have the new, energy efficient stoves; a market should be in Baihualing hamlet near the Village Office instead.
Manggang	Skepticism about benefits from tourism since tourists only go to the Ranger Station; people could sell wild mushrooms and other foods; hopeful tourism will bring economic benefits to all.
Manghuang	Skepticism about any benefits, hope for tourists to visit local stores, people need more cash to invest in anything; a market should be in Hanlong.

生态旅游者感兴趣的植物 PLANTS OF INTEREST TO ECOTOURISTS

种 Species	科 Family	利用 Uses	说明 Notes
Pistacia weinmannifolia J. Poisson ex Franch.	Anacardiaceae	C	仅见于高黎贡山 Only in S Gaoligongshan
Amorphophallus spp.	Araceae	F	根富含淀粉 Rhizomes rich in starch
Eleutherococcus gracistylis (W. W. Smith) S. Y. Hu	Araliaceae	M	对治疗关节炎有疗效 arthritis
Hedera sinensis (Tobler) Hand.-Mazz. Common climber with	Araliaceae	O	中国长春藤，一种常见攀缘植物， 具常绿叶 Common climber with evergreen leaves; Chinese ivy
Panax variabilis J. Wen	Araliaceae	M	对治疗关节炎有疗效 arthritis
Panax bipinnatifidus Seem.	Araliaceae	M	对治疗关节炎有疗效 arthritis
Merrilliopanax listeri (King) Li	Araliaceae	O	灌木，喜阴凉环境 Shrub at shady sites
Macropanax dispermus (Bl.) Kuntze	Araliaceae	O	乔木，叶呈掌状，极具观赏价值 Tree with beautiful palmate leaves
Brassaiopsis fatsioides Harms	Araliaceae	O	灌木，具大的掌状叶 Shrub with unique and large palmate leaves
Brassaiopsis palmipes Forrest ex W. W. Smith	Araliaceae	O	灌木，具大的掌状叶 Shrub with unique and large palmate leaves
Metapanax delavayi (Franch.) J. Wen & Frodin	Araliaceae	O, M	对治疗关节炎有疗效 arthritis
Schefflera shweliensis W. W. Smith	Araliaceae	O	依生境不同或为优雅的乔木或 为漂亮的灌木 Elegant trees and shrubs at various habitats
Begonia clavicaulis Irmsch.	Begoniaceae	O	草本植物 Herb
Begonia forrestii Irmsch.	Begoniaceae	O	草本植物 Herb
Mahonia polyodonta Fedde	Berberidaceae	O, M	灌木，具美丽花和叶 Shrub with beautiful leaves & flowers
Mahonia paucijuga C. Y. Wu	Berberidaceae	O, M	
Sambucus adnata Wall.	Caprifoliaceae	O	大型草本 Large herb
Sambucus chinensis Lindl.	Caprifoliaceae	O, M	疗伤药 Wounds
Viburnum cylindricum Buch.-Ham. ex D. Don	Caprifoliaceae	O, M	活血 Blood -regulating
Nyssa shweliensis (W. W. Smith) Airy-Shaw	Cornaceae	C	仅分布于云南省藤冲 Rare & local in Tengchong, Yunnan
Gynostemma pentaphyllum (Thunb.) Makino	Curcurbitaceae	M	用途广泛 Used popularly
Alsophila spinulosa Wall. ex (Hook.) Tryon	Cyatheaceae	C	树蕨，国家一级保护物种 Tree fern, Class I endangered in China
Elaeocarpus lanceaefolius Roxb.	Elaeocarpaceae	O	乔木，具大而可食用的果实 Tree with large edible fruits
Rhododendron decorum Franch.	Ericaceae	O	具绚丽的白花 Showy white flowers
Rhododendron delavayi Franch.	Ericaceae	O	–
Rhododendron neriiflorum Franch.	Ericaceae	O	中、高海拔常见，开艳丽的红花 Common at mid to high altitudes with showy red flowers
Rhododendron protistum Balf. F. & Forrest var. *giganteum* (Tagg.) Chamb.	Ericaceae	O	也称杜鹃王 Also known as the king of rhododendrons

Uses

C = 保护 Conservation	M = 药用 Medicinal	T = 用材 Timber
F = 食品 Food	O = 观赏 Ornamental	

Plants of potential interest to ecotourists found in the Gaoligong Mountain National Nature Reserve, Yunnan, China. List prepared by Jun Wen.

生态旅游者感兴趣的植物 PLANTS OF INTEREST TO ECOTOURISTS

种 Species	科 Family	利用 Uses	说明 Notes
Rhododendron simsii Planch.	Ericaceae	O	灌木，一般开艳丽的红花 Shrub, common with showy red flowers
Agapetes lacei Craib.	Ericaceae	O	常绿灌木，冬季开花 Evergreen shrub bearing winter flowers
Eucommia ulmoides Oliv.	Eucommiaceae	O, M	常用于治疗高血压 Commonly used to cure high blood pressure
Euptelea pleiosperma Hooker F. & Thoms.	Eupteleaceae	O	–
Cassia sophera L.	Fabaceae	O	小灌木 Small shrub
Albizia odoratissima (L.) Benth.	Fabaceae	O	一种乔木，多花，常见于开阔地上 Common tree with showy flowers at open sites
Gentiana intricata Marq.	Gentianaceae	O	–
Exbucklandia populnea (R. Br. ex Griff.) R. Br.	Hamamelidaceae	O, T	–
Hydrangea chinensis Maxim.	Hydrangeaceae	O	具艳丽的花 Showy flowers
Hypericum spp.	Hypericaceae	M	–
Iris spp.	Iridaceae	O	高黎贡山约有10种 ca. 10 spp. In Gaoligongshan
Lilium spp.	Liliaceae	O	在高黎贡山有几种分布 a few species
Ophiopogon bodinieri Levl.	Liliaceae	O	草本植物 Herb
Fritillaria cirrhosa D. Don	Liliaceae	M	对治疗感冒和咳嗽有疗效 Cold & cough
Buddleja asiatica Lour.	Loganiaceae	O	灌木 Shrub
Manglietia insignis (Wall.) Bl.	Magnoliaceae	O	乔木，树形优美，具绚丽的花 Elegant tree with showy flowers
Manglietia hookeri Cubitt. & W. W. Smith	Magnoliaceae	O	乔木，树形优美，具绚丽的花 Elegant tree with showy flowers
Magnolia rostrata W. W. Smith	Magnoliaceae	O	–
Magnolia nitida W. W. Smith	Magnoliaceae	O	–
Michelia floribunda Fin. & Gagnep.	Magnoliaceae	O	–
Ardisia maculosa Mez	Myrsinaceae	O	冬春季结红色的果实 Red fruits in winter and spring
Embelia floribunda Wall.	Myrsinaceae	O	果实成熟时呈鲜红色，从冬季一直延续到初夏 Bright red fruits in winter into early summer
Nelumbo nucifera Gaertn.	Nelumbonaceae	O	水生植物，具艳丽的花及盾状的叶 Aquatic plant with showy flowers and unique peltate leaves
Epilobium amurense Hausskn.	Onagraceae	O	–
Dendrobium spp.	Orchidaceae	O	–
Calanthe spp.	Orchidaceae	O	–
Gastrodia elata Bl.	Orchidaceae	C, M	国家三级保护物种 Class III endangered in China
Peperomia dindygulensis Miq.	Piperaceae	O	草本植物 Herb
Peperomia tetraphylla (Forst. F.) Hooker f. & Arn.	Piperaceae	O	–
Piper macropodum DC.	Piperaceae	O	–

Uses

C = 保护 Conservation   M = 药用 Medicinal   T = 用材 Timber
F = 食品 Food   O = 观赏 Ornamental

生态旅游者感兴趣的植物　PLANTS OF INTEREST TO ECOTOURISTS			
种 Species	科 Family	利用 Uses	说明 Notes
Piper pedicellatum DC.	Piperaceae	O	攀缘植物　Climber
Polygala didyma C. Y. Wu ex C. Y. Wu & S. K. Chen	Polygalaceae	O	–
Polygala arillata Buch.-Ham. ex D. Don	Polygalaceae	O	–
Primula spp.	Primulaceae	O	–
Helicia shweliensis W. W. Smith	Proteaceae	C, O	国家三级保护物种 Class III endangered in China
Caltha palustris L.	Ranunculaceae	O	常见草本植物，颜色艳丽，分布于 高海拔地区 Showy & common herb at higher altitudes
Clematis nepaulensis DC.	Ranunculaceae	O	攀缘植物，见于中海拔地带 Climber at mid altitudes
Coptis teata Wall.	Ranunculaceae	C, M	中国二级保护树种 Class II endangered in China
Prunus cerasoides D. Don	Rosaceae	O	花美丽，果可食，常见于中海拔地带 Showy flowers & edible fruits, common at mid altitudes
Rubus lineatus Reinw.	Rosaceae	O	灌木，多见开阔地 Shrubs at open sites
Rosa odorata (Andr.) Sweet	Rosaceae	C, O	国家三级保护物种，常见于腾冲和保山 Class III endangered in China, but frequent in Tengchong & Baoshan
Mussaenda erosa Champ.	Rubiaceae	O	灌木，具艳丽花序 Shrub with showy flowers
Luculia intermedia Hutch.	Rubiaceae	O, M	活血，治疗蛇咬伤 Blood regulating & snake bites
Wendlandia speciosa Cowan	Rubiaceae	O	乔木或灌木 Trees & shrubs
Pyrularia edulis (Wall.) DC	Santalaceae	O	常见，可提取油料 Commonly used to extract oil
Pedicularis spp.	Scrophulariaceae	O	在高黎贡山高山地带有多种分布 High species diversity in Gaoligonshan high altitudes
Sinopteris grevilleoides (Christ.) C. Chr. & Ching	Sinopteridaceae	C	中国二级保护树种 Class II endangered in China
Stachyurus chinensis Franch.	Stachyuraceae	O	–
Staphylea shweliensis W. W. Smith	Staphyleaceae	C	仅分布于云南省腾冲 Only in Tengchong, Yunnan
Taxus yunnanensis Cheng & L. K. Fu	Taxaceae	O	灌木或小乔木，树形优美 Elegant shady shrub or small tree
Taiwania flousiana Gaussen	Taxodiaceae	O	树形优美，中国大陆和台湾间断分布 Beautiful tree, with a disjunct distribution between mainland China and Taiwan
Tetracentron sinense Oliver	Tetracentraceae	O	–
Camellia japonica L.	Theaceae	O	灌木，具美丽花和叶 Shrub with beautiful leaves & flowers
Camellia sinensis (L.) O. Ktxe.	Theaceae	F	茶树 Chinese tea

Uses

C = 保护 Conservation　　　　M = 药用 Medicinal　　　　T = 用材 Timber

F = 食品 Food　　　　O = 观赏 Ornamental

Plants of potential interest to ecotourists found in the Gaoligong Mountain National Nature Reserve, Yunnan, China. List prepared by Jun Wen.

生态旅游者感兴趣的植物 PLANTS OF INTEREST TO ECOTOURISTS

种 Species	科 Family	利用 Uses	说明 Notes
Schima wallichii (DC) Korth.	Theaceae	O	常见乔木，开大白花 Common tree with large flowers
Gordonia longicarpa H. T. Chang	Theaceae	O	–
Daphne papyracea Wall.	Thymelaeaceae	O	一种灌木 Shrub
Edgeworthia gardneri (Wall.) Meissn.	Thymelaeaceae	O	–
Paris spp.	Trilliaceae	M	治疗蛇咬伤，止血 Snake bites, stopping bleeding

Uses

C = 保护 Conservation	M = 药用 Medicinal	T = 用材 Timber
F = 食品 Food	O = 观赏 Ornamental	

生态旅舍和游客接待中心设计

百花岭生态旅游人口区

我们用"人口区"一词来指紧挨着目前的保护区管理站周围的地区。该地区有公路可达，距离保护区边界约500m左右。该区域重要的旅游设施和景点包括管理站、生态旅舍、游客接待中心、古南方丝绸之路和几个民族村寨等。

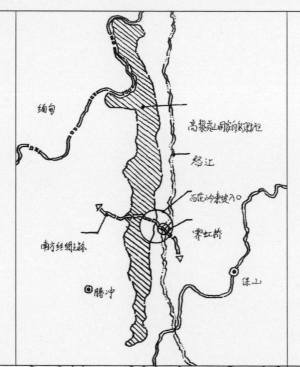

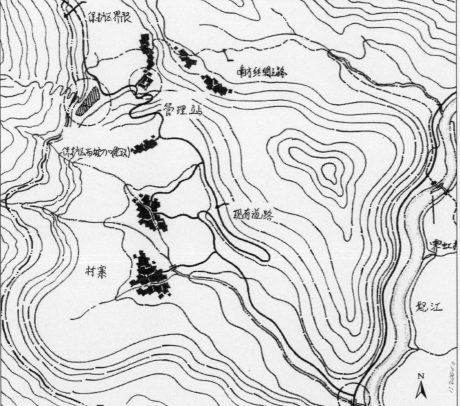

该地区最有特色的历史古迹当属"古南方丝绸之路"。古道在这一带保存较好，石头铺就的路从林子和村中穿过。

目前的管理站有些供游客和科学考察人员住宿的房间，还有一个简易食堂和洗浴室。管理站的布局类似一个四合院，院子中央有一个小的喷泉和一些花草，院子里还可以停车。管理站位于汉龙村地界。

生态旅舍和游客接待中心设计

百花岭生态旅游入口区

建议	01 尽量利用现有的基础设施，充分发挥新修设施和管理所现有设施的作用
	02 修建连接生态旅舍、游客接待中心、南方丝绸之路、管理所和自然村等的步行便道图

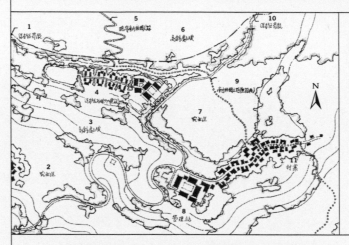

图题游客接待中心的规模、布局等应与当地汉龙村的传统村寨的风格一致。

03 在旅舍区和管理所修建专门停车场，以免损害环境敏感区域？

04 对南方丝绸古道进行研究，确定适合修建步行便道的地段

05 利用多语种编写介绍南方古丝绸之路的文化和自然历史的材料；
制定标牌，规范游客行为。

06 将现在的管理所的设施作为旅游接待的辅助设施，或用于开展教育培训或科研活动

07 对管理所的入口处进行重新设计，以加强游客的视觉效果

08 制定涵盖整个入口村（包括自然村）的垃圾管理计划。有些垃圾可以在指定地点进行
堆放做堆肥；定期收集可回收利用的垃圾，运走进行处理。在设计处理和堆放场地的时
候一定要小心以尽量避免臭味四溢和影响附近的卫生条件

09 开展研究，了解入口区和游客接待中心的化粪池对周围地区的影响

10 配置备用发电机，准备整个入口区都可以利用的应急资源

11 为入口区所有设施寻求替代能源，包括光伏电池、风力发电机和太阳能热水器等

生态旅舍和游客接待中心设计

百花岭入口区生态旅舍和游客接待中心

我们建议在保护区外和现在的管理站附近的区域修建供游客食宿和对游客进行教育的设施。该设施一方面为到这一带旅游的游客提供食宿，又可以给保护区周围其它设施的修建起到样板的作用。

由于该地区生态上的敏感性，设施的规模应该以每晚接待游客不超过100人为准。设计的原则是尽量不要对修建地点、保护区和周边社区带来负面影响。除了生态环境方面的考虑外，小规模的开发也有利于市场销售，因为高端游客一般喜欢比较亲切、比较个性化的旅游体验，更喜欢规模比较小的旅舍。

设计和建设生态旅舍和游客接待中心应遵循下列原则：

1) 利用已有设施

2) 利用符合当地传统的建筑材料或其来源不会给当地环境造成损害的材料

3) 严格限制和消除对修建地点带来的消极影响，如避免造成水土流失、侵占耕地和对水系造成扰动等

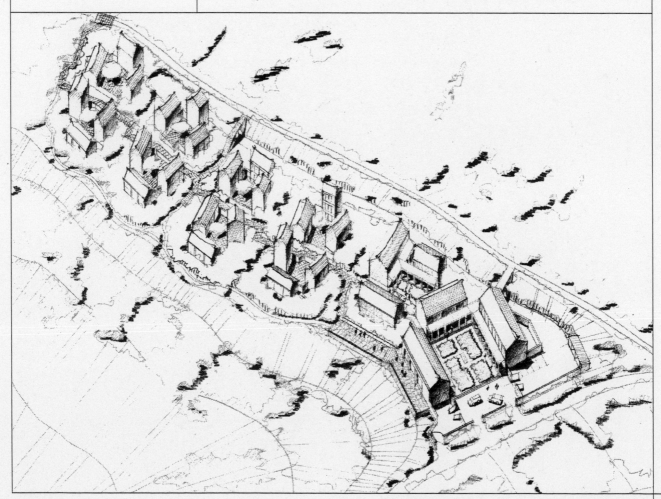

生态旅舍和游客接待中心设计

百花岭人口区生态旅舍和游客接待中心

建议	进入生态旅游区
	01 一切人员都从现在已有的道路进入旅舍和游客接待中心,不用再新修公路。
	02 设计旅游活动时,要对车辆等进行严格控制,游客不能自驾车辆进入旅游区。
	03 对现在通往百花岭的路的路面进行平整,但是要注意保存其乡村特色; 路面不要再加宽。
	04 在旅舍附近修建适当的(规模不宜太大)停车场
	05 修建道路,将旅舍和附近的旅游景点的人行便道等连接起来。
	建筑区规划
	01 各建筑物应该高低错落有致,但是最高不要超过两层或9m高。
	02 旅舍和游客接待中心的布局要自然,要能够反映当地村民庭院设计的特点和传统 村寨的建筑特征。
	03 旅舍应该具有良好的采光和通风条件。
	04 要充分考虑各个房间的视野,使游客从所有的房间都能欣赏到外面的田园风光 或山水景致。
	05 尽量将服务区设计在游览区以外或从游览区看不到的区域。

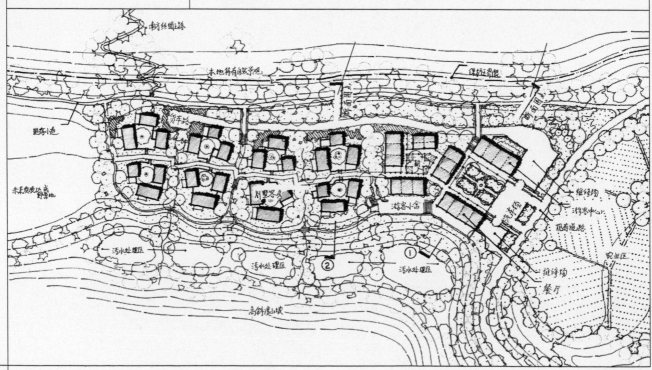

生态旅舍和游客接待中心设计

百花岭入口区生态旅舍和游客接待中心

游客住宿区剖面图

生态旅舍和游客接待中心设计

百花岭入口区生态旅舍和游客接待中心

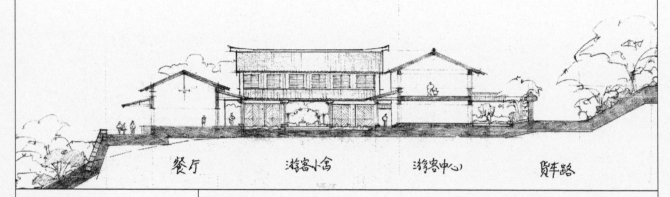

餐厅　　　　游客小舍　　　游客中心　　　货车路

建议	建筑设计
	01 利用当地传统的建筑设计思路，新修的房屋要与周围村寨的房屋风格（包括屋顶线、材料等）协调一致
	02 利用当地人用的建筑材料如石头、沙和砾石等。注意采石地点应该避开环境敏感区域（石料在当地并不缺乏）。
	03 利用当地的沙石做骨料来配制混凝土，从而产生具有当地特色的颜色和饰纹
	04 有选择地使用木材，以避免过度消耗森林资源。木材只应该用在人体需要接触的地方或用在少数需要修饰的地方。
	05 可以尽量使用当地比较丰富的竹子资源。竹子的用途广泛，可以用做地板、家具、装饰等。
	06 利用钢筋做混凝土模板，模板以后还可以循环使用
	07 设计能够反映当地式样、色彩和饰纹的装饰风格。不过，不能简单地模仿传统建筑结构，这样反而会影响其真实性。

生态旅舍和游客接待中心设计

百花岭入口区生态旅舍和游客接待中心

建议	可持续原则
	01 所有建筑设施包括游客接待设施和工作人员生活设施等都要做到最大程度地利用日光. 白天尽量不要用电, 只在夜晚或在特殊情况下才用电
	02 在不对脆弱的生态造成损害的情况下使用当地材料
	03 利用建设游客住宿设施的机会, 对当地人进行培训. 应该对工人们进行设计原则和建筑技巧等方面的培训
	04 在需要使用木材的时候, 一定要保证使用的木材来源合法, 或从现存的可回收资源中获取
	05 通过吊扇或正确设计窗户等方式来提高房屋的通风水平, 不使用空调装置.
	06 对游客进行正确使用可持续设施方面的教育以及当地文化和传统的教育

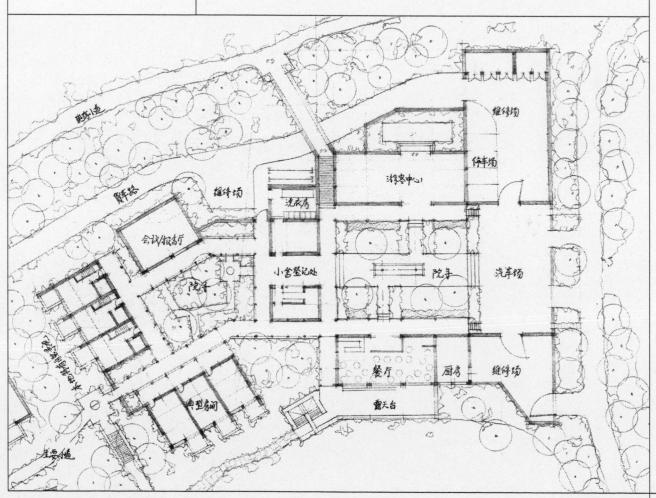

Design of the Baihualing Gateway Complex for ecotourism in the eastern slope of Gaoligongshan.
Particpants: Peter J. Kindel, Adam Thies.

GATEWAY LODGE AND VISITOR CENTER DESIGN

The Baihualing Gateway Complex

We use the term Gateway Complex to describe the area of Baihualing immediately surrounding the existing ranger station. This area is served by one roadway and is situated approximately 500 meters from the current reserve boundary. Important features of the complex include the ranger station, the new Gateway Lodge and Visitor Center site, the Southern Silk Road and several hamlets.

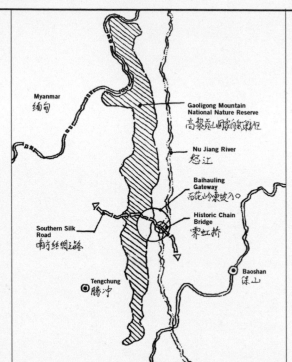

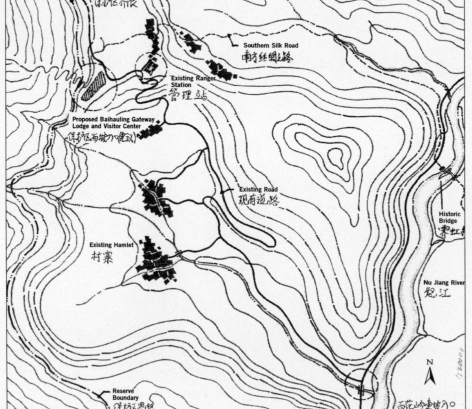

A unique historical feature in the Gateway Complex area is the Southern Silk Road. The trail is pronounced in this area, with ancient stonework forming a pathway through the forest and villages.

The existing ranger station contains dormitory-style rooms for visiting officials and scientists, as well as basic food service facilities and bathrooms. The entire facility surrounds a courtyard with a small fountain, parking areas and vegetation. Adjacent to the ranger station is the hamlet known as Hanlong.

GATEWAY LODGE AND VISITOR CENTER DESIGN

The Baihualing Gateway Complex

Recommendations

01 Build on existing infrastructure to the extent possible, and create efficiencies of use between the new lodge and the existing ranger station in both facilities and programming.

02 Establish clear pedestrian links between the Gateway Lodge and Visitor Center, Southern Silk Road, ranger station and local hamlets.

1 - Reserve Boundary
2 - Agricultural Area
3 - Steep Slopes
4 - Proposed Baihualing Gateway Lodge and Visitor Center
5 - Southern Silk Road Route (extant)
6 - Steep Slopes
7 - Agricultural Area
8 - Existing Ranger Station
9 - Southern Silk Road Route (no longer visible)
10 - Reserve Boundary
11 - Existing hamlet of Hanlong

The size, massing and siting of the new Lodge and Visitor Center should be similar to traditional villages, like the nearby hamlet of Hanlong.

03 Establish defined parking areas within the ranger station and lodge for vehicle parking and storage to avoid damaging sensitive environmental areas.

04 Study the Southern Silk Road to determine which sections are appropriate as pedestrian trails.

05 Develop written multi-lingual materials that document the cultural and natural history of the Southern Silk Road. Develop signage to inform visitors of appropriate usage and trail activities.

06 Maintain the ranger station as a support facility for park activities and to house occasional education events and research activities. (Allow low-budget travelers to stay in the lodging facilities when available.)

07 Redesign the entrance to the ranger station to create a strong visual experience as the ecotourist continues to the Gateway Lodge and Visitor Center.

08 Develop careful waste management plans that benefit *all* users in the Gateway Complex, including local hamlets. Waste that is natural in content should be composted for agricultural uses on designated sites. Collect recyclable waste materials on a regular basis and remove them for processing. Design the processing and holding site with utmost care to avoid creating an unsightly facility with poor odors and unsanitary conditions.

09 Conduct a careful study to understand the impact of the new Gateway Lodge and Visitor Center's additional septic tank leeching to the Gateway Complex area.

10 Develop emergency generators and shared emergency resources for the benefit of the entire gateway complex.

11 Examine options for sustainable energy alternatives for all Gateway Complex facilities. These alternatives might include photovoltaic cells, wind generated turbines, and solar power and heating.

GATEWAY LODGE AND VISITOR CENTER DESIGN

The Baihualing Gateway Lodge and Visitor Center

We propose the development of a facility to provide lodging, dining and visitor orientation adjacent to the reserve and to the existing ranger station. This facility will accommodate visitors to this portion of the reserve, and will serve as a prototype for all facility development around the reserve.

Due to the ecological sensitivity of the site, this facility should be limited to a maximum of 100 guests per night. It must be designed according to principles that will minimize its impact on the site, the reserve, and all neighboring hamlets. A smaller-scale development is also beneficial from a marketing point of view, because high-end tourists typically seek an intimate, personalized experience and prefer a small facility to a large hotel. Guidelines for the design and construction of the lodge and visitor center include:

> Use existing infrastructure to access and service the facility.

> Use materials that are consistent with local traditions and are obtained without negative impact to the environment.

> Control and eliminate negative impacts to the site including erosion, excessive grading, forest clearing, loss of agricultural land, and hydrologic interruptions.

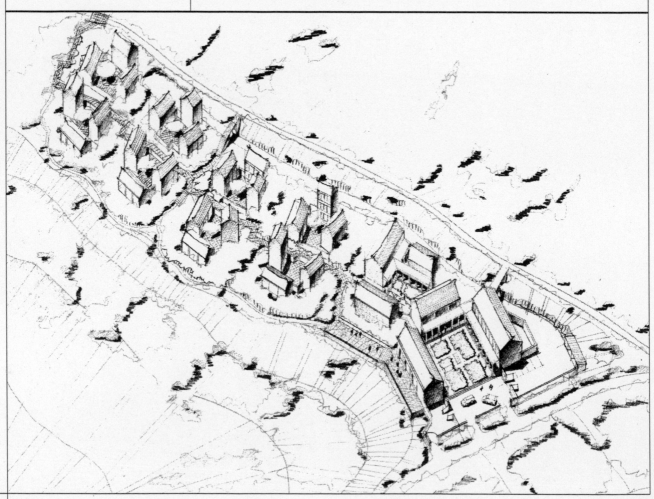

GATEWAY LODGE AND VISITOR CENTER DESIGN

The Baihualing Gateway Lodge and Visitor Center

Recommendations	

ACCESS

01 Provide all access to the site via the existing road. Do not construct new or additional roads to service the Gateway Lodge and Visitor Center.

02 Plan for visits that include controlled transportation – visitors will not be driving their own vehicles.

03 Improve the surface of the existing access road while preserving its rustic quality. Do not widen the road.

04 Provide limited parking for jitneys and vans at the lodge.

05 Create direct links between trails and pedestrian routes within the lodge and the surrounding attractions.

SITE PLAN

01 Design structures of varied height, but never exceed 2 levels, or 9 meters to peak of roof.

02 Create an informal layout of Lodge and Visitor Center buildings that reflects the court-yard house design and informal structure of traditional villages.

03 Expose all four walls of guest accommodations to light and ventilation.

04 Provide direct views or connections to landscaped areas in all structures. Design all guest units to provide views to the surrounding mountains or valleys.

05 Design discreet service areas that are not in direct view of public areas.

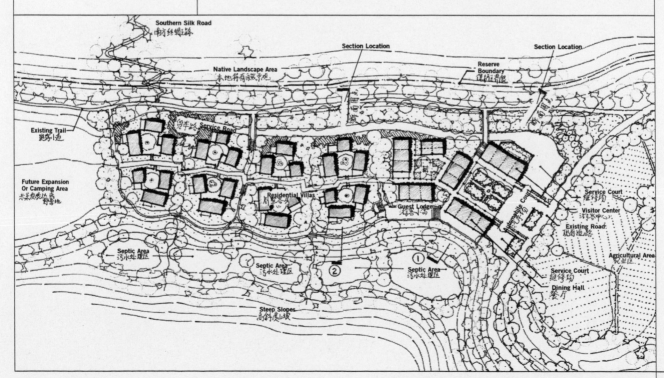

GATEWAY LODGE AND VISITOR CENTER DESIGN

The Baihualing Gateway Lodge and Visitor Center

Cross-section through guest units

Plan of guest units

GATEWAY LODGE AND VISITOR CENTER DESIGN

The Baihualing Gateway Lodge and Visitor Center

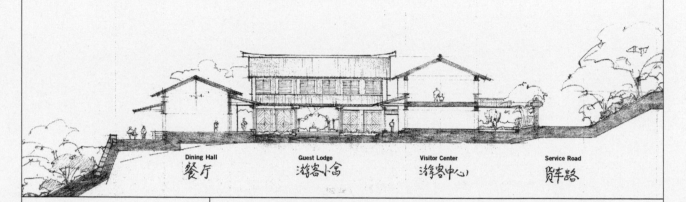

| Dining Hall | Guest Lodge | Visitor Center | Service Road |
| 餐厅 | 游客小舍 | 游客中心 | 贤车路 |

Recommendations

ARCHITECTURAL DESIGN

01 Use interpretations of indigenous architectural traditions, and design buildings to harmonize in massing, roof-lines, and materials with surrounding villages.

02 Employ indigenous materials, including stone, sand and gravel. Note that quarrying should not take place in sensitive areas around the reserve. Local stone is available throughout the Baoshan region.

03 Mix concrete using local sand and gravel, with a strong aggregate additive to generate color and texture.

04 Use wood very selectively in construction to prevent further deforestation. It should be limited to areas of human contact, or in modest decorative elements.

05 Use bamboo, which is widely available in the area, throughout the project. This versatile material should be used for flooring, cabinetry, furnishings, decorative elements, scaffolding, and other purposes.

06 Explore use of steel for concrete forms, which can be recycled for future use.

07 Design building detailing that strongly reflects local patterns, color and texture. Detailing should not, however, mimic or recreate traditional structures—this would compromise authenticity.

GATEWAY LODGE AND VISITOR CENTER DESIGN

The Baihualing Gateway Lodge and Visitor Center

Recommendations	SUSTAINABLE PRINCIPLES
	01 Provide maximum daylight to all structures and facilities, including service and employee areas. Use electric lighting only at night, or in special conditions.
	02 Employ local materials without harming the nearby fragile ecology.
	03 Use the construction of the lodge as an opportunity to provide skilled training for surrounding populace. Workers should be trained in design principles, as well as in building trades and skills.
	04 Obtain wood either from certified, managed forests, or recycle it from other appropriate existing sources of recyclable materials.
	05 Provide ventilation with ceiling fans and fully operational windows. Air conditioning will <u>not</u> be provided—instead, passive ventilation measures will be encouraged. Additional cooling will be provided by the thermal mass of the structures.
	06 Educate visitors about the operation and appropriate use of a sustainable facility, as well about local cultures and traditions.

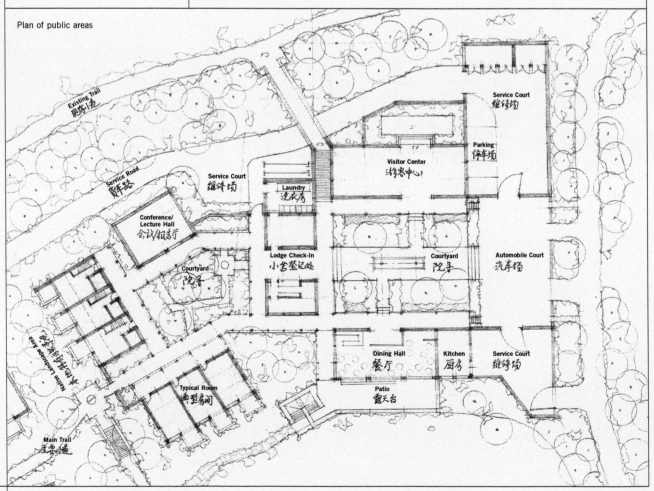

Plan of public areas

高黎贡山国家级自然保护区保山管理局年度报告（2002）

Center for U.S.-China Arts Exchange; Skidmore, Owings & Merrill; and Openlands Project. 2001. The Weishan Heritage Valley; Recommendations for preservation and future growth. New York: Center for U.S.-China Arts Exchange.

Collar, N. J., M. J. Crosby, and A. J. Stattersfield. 1994. Birds to Watch 2: The World List of Threatened Birds. Birdlife International, Cambridge, UK.

李恒等. 2000. 高黎贡山植物. 北京:科学出版社

MacKinnon, J. and K. Phillipps. 2000. A Field Guide to the Birds of China. Oxford University Press, Oxford, UK.

云南省人口普查办公室. 1992. 1990年云南省人口普查. 北京:中国统计出版社

Stattersfield, A. J., M. J. Crosby, A. J. Long, and D. C. Wege. 1998. Endemic Bird Areas of the World: Priorities for Biodiversity Conservation. Birdlife International, Cambridge, UK.

薛纪如等. 1995. 高黎贡山国家自然保护区. 北京:中国林业出版社

Baoshan Management District, 2002. Annual Report.

Center for U.S.-China Arts Exchange; Skidmore, Owings & Merrill; and Openlands Project. 2001. The Weishan Heritage Valley; Recommendations for preservation and future growth. New York: Center for U.S.-China Arts Exchange.

Collar, N.J., M.J. Crosby, and A.J. Stattersfield. 1994. Birds to Watch 2: The World List of Threatened Birds. Birdlife International, Cambridge, UK.

Li, H., H. Guo, and Z. Bao. 2000. Flora of Gaoligong Mountains. Science Press, Beijing, China.

MacKinnon, J. and K. Phillipps. 2000. A Field Guide to the Birds of China. Oxford University Press, Oxford, UK.

Population Census Office of Yunnan Province, 1992. 1990 Population Census of Yunnan Province, China Statistical Publishing House.

Stattersfield, A.J., M.J. Crosby, A.J. Long, and D.C. Wege. 1998. Endemic Bird Areas of the World: Priorities for Biodiversity Conservation. Birdlife International, Cambridge, UK.

Xue, J.R. et. al. (Ed.). 1995. Gaoligong Mountain National Nature Reserve (in Chinese). China Forestry Publishing House, Beijing, China.

Alverson, W. S., D. K. Moskovits, and J. M. Shopland (eds.).
2000. Bolivia: Pando, Río Tahuamanu. Rapid Biological
Inventories Report 01. Chicago: The Field Museum.

Alverson, W. S., L. O. Rodríguez, and D. K. Moskovits (eds.). 2001.
Perú: Biabo Cordillera Azul. Rapid Biological Inventories
Report 02. Chicago: The Field Museum.

Pitman, N., D. K. Moskovits, W. S. Alverson, and
R. Borman A. (eds.). 2002. Ecuador: Serranías Cofán–
Bermejo, Sinangoe. Rapid Biological Inventories
Report 03. Chicago: The Field Museum.